THE
SCIENCE
OF
COMPOST

www.pimpernelpress.com

The Science of Compost: Life, Death & Decay in the Garden
© Pimpernel Press Limited 2022
Text © Julian Doberski 2022
Illustrations © Pimpernel Press Ltd 2022
Except as noted on page 111

A catalogue record for this book is available from
the British Library

Designed by Sarah Pyke
Typeset in Josefin Sans

ISBN 978-1-914902-93-2
Printed and bound in Great Britain

The author has made every effort to contact holders of
copyright works. Any copyright holders he has been
unable to reach are invited to contact the author
or the publisher so that a full acknowledgement
may be given in subsequent editions.

9 8 7 6 5 4 3 2 1

Julian Doberski

THE
SCIENCE
OF
COMPOST

Life, Death + Decay in the Garden

Pimpernel
Press ltd
www.pimpernelpress.com

CONTENTS

1 / INTRODUCTION

Decay - the destruction of life's organic matter - is the core process in compost and composting. Really, composting is part of the same 'natural' decay processes that help to drive the living world. In fact, around 80 to 90 per cent of plant growth (leaves, stems, flowers and so on) will go through decay, rather than being eaten by plant-eating animals (herbivores). And even herbivores produce faeces derived from plant material, which must also decay.

In woods and forests, the need for decay is obvious - especially in temperate regions. Here, during the autumn months, plants die or shed large quantities of leaf or needle litter. Typically, extensive decay will then follow so that by the next summer we can see the evidence of change, with only a thin layer of leaf debris remaining. In some other woodlands, depending on the tree species, there may still be a thick, spongy layer of only partially decomposed leaf material. So already we can see that although decay is inevitable, not all plant organic material is equal in terms of decomposition. Decay can be fast, or it can be slow.

The processes that decay both woodland leaf litter and garden compost are largely hidden, certainly out of obvious sight. But without those processes, the materials on which life depends would simply accumulate. Without nutrient recycling and new growth, the living world would slowly wind down and eventually stop. However, composting is also different from decay in the natural world. The complexity of this will be explored further later but basically involves human intervention: in composting, the organic matter is aggregated in some way and the composition manipulated in ways that do not happen in nature. There can be further intervention too - typically to make composting faster and more complete.

Increasingly, people are using the power of composting to help get rid of the imbalances in natural recycling that they are so good at creating. Whether it is the grass cut from lawns that must not spoil the pristine appearance of our gardens, the skins from bananas that have travelled halfway across the globe to feed us, or indeed many other biologically derived materials – that decay process needs to happen. This can be slow in a waste landfill site or quick in a specialized composting facility, where conditions are optimized to speed up the process. In this sense composting is rather like cooking: there is skill in balancing and refining the processes involved to quickly achieve an excellent product. But this book is not about the practicalities of running a commercial composting plant, nor does it provide a 'recipe' for building the perfect garden compost heap. Rather it is about starting to understand what goes on in the compost heap or bin in the corner of your garden. Compost is not just a pile of organic matter. Instead, it can be thought of as a largely isolated but diverse, transient and dynamic ecosystem. The question is, how does that ecosystem work?

Usually, a heap of compost is seen only from the outside. To understand what is really going on, we must delve into the interior of the compost material. You will likely get a hint of this when you dig out the heap to remove the (hopefully) friable compost to serve as mulch, soil conditioner or fertilizer in the garden. As you dig, you may casually notice worms sliding away and other creatures scuttling back into the heap's dark interior. But let us go back to a point just before you dismantle the heap. Imagine being shrunk to a fraction of your normal size and standing at the entrance of a compost 'cave system'. You enter through a gap in the pile of leaves and organic debris and start exploring like a potholer, switching on a torch as you work your way through the compost. Unlike a limestone cavern, the very walls around you are being eaten or rotted away by bacteria, fungi and an array of invertebrates feeding on the plant detritus or the microbes or each other. As you shine your tiny light on this frenetic scene, the tangled webs of fungal strands, the glistening colonies of bacteria and the monstrous-looking insects give a glimpse into an active world of life – a precursor to death and essential for decay to happen. This is a dynamic living world that changes with every passing day.

So, our mind game offers a vision of the complexity of the compost ecosystem. It is not one based directly on photosynthesizing plants, as is true in the world of sunshine, but is instead an ecosystem based on energy stored in the organic detritus that feeds the compost heap. While not separate from a sunshine-driven world, it is a subsystem that we can think of as largely self-contained. Parallels have been drawn between the ecological complexity of compost and that of a coral reef, which is also a diverse and complex community distinct from others around it. This may seem like a somewhat fanciful suggestion, but the reality of the comparison becomes more apparent when you step down to the scale shown by a magnifying lens or microscope. With magnification it becomes clear that compost is heaving with diverse life, with diverse functions. It would be impossible to fully dissect and explain this complexity in a short book, so inevitably the picture presented here will be partial. Nevertheless, this exploration should provide a better understanding of what goes on in that pile in the corner of the garden. It will also point to some of the numerous known unknowns!

The basic elements of the input, recycling and then production of new organic matter in a compost heap are shown in Figure 1. The process is essentially one of oxidation of organic matter, which leaves a residue of tougher, slowly decomposing material referred to as compost. But, of course, that process itself sustains the growth of new life within the compost, be it microbial or animal. In the following chapters we will be discussing the different processes shown in the diagram and you may find it useful to occasionally flick back to this figure to locate where we are in the cycle and to understand it more fully.

Obviously, the limitations of current technology mean you can't actually shrink yourself down to the right size for compost heap potholing, so our scenario requires some imagination. Nevertheless, the world of natural history film-making has occasionally provided a pictorial introduction to the complexities of this world. There are also relatively cheap (non-medical) endoscopes available that can be linked to a

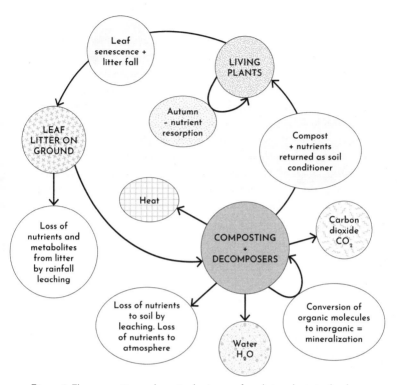

Figure 1. The composting cycle: a circular journey from living plants to dead plant organic matter to decomposing compost material – before the return of compost with nutrients to the soil to supply the next seasonal upsurge in plant growth. Loss of compost volume is by decomposition of cellulose and other materials and conversion to carbon dioxide and water.

mobile phone for a degree of close-up exploration. But there is certainly one location where you don't need to be small to access and appreciate some of the dynamic aspects of decomposition. The Gomantong Caves on the island of Borneo have been made famous by television footage by Sir David Attenborough and his team. This cave system is home to millions of bats and thousands of swiftlets, which between them generate a huge pile of guano (faeces) on the floor of the cave. This is an example of a detritus-based ecosystem, albeit with animal rather than plant organic waste. Unlike a deceptively quiet compost heap, the surface here is a seething mass of cockroaches, oversized, long-legged centipedes and a host of other less-visible invertebrates, bacteria and fungi, consuming or decomposing the guano, dead bats and swiftlets, and with invertebrates eating each other. Having been there myself, I can vouch for the fact that this is probably the closest you can get to the experience of being 'in' a kind of compost heap. It provides a powerful vision of a detritus- and decomposition-based ecosystem in action and the kind of activity also evident (though less dramatically so) when you get up close and personal with your compost.

DECAY –
THE
DESTRUCTION
OF LIFE'S
ORGANIC
MATTER

2 / WHAT IS COMPOST?

In a sense, everyone has an idea of what is meant by compost. Many of us have at the bottom of the garden an untidy and rather brooding pile we (more or less) loosely refer to as compost. But let's try to be a bit more exact. A good start is a definition of the word. One dictionary defines compost as 'Decaying plant material that is added to soil to improve its quality'. We have briefly touched upon the fact that decay is the core process in compost, but this definition rather lacks the essence of what makes compost different from leaves decaying on the forest floor. In the garden, composting involves 'bulking up' dead plant material and often mixing it with other biologically derived materials – not necessarily all of plant origin. This is what most gardeners think of as a 'compost heap', or else the contents of a more organized compost bin of some kind. Beyond this garden scale, composting enters the world of business – where commercial processing requires large-scale industrial machinery for the quickest throughput of tons of compost, leaving residue in the form of fully composted and usable material.

That simple dictionary definition fails to capture the spectrum of meanings to the word compost. It can refer to a heap of organic material just starting out on the process of decay, or it can equally well be used to describe the material at the end point of decay when the original decaying organic matter is no longer clearly recognizable. It is this desired end point to composting that is a gardener's delight – a wonderful amorphous material to dig in as a soil conditioner.

As with other words of biological origin, the meaning of the word compost has been extended beyond its original use. Bags of compost purchased in a garden centre may be derived from industrial-scale composting of, for example, material from domestic 'green' bins. But compost may also refer to peat from ancient deposits of partially decomposed plant matter mined for horticultural use. This kind of peat-based compost will not be considered here. The focus will be on active composting of organic matter (primarily plant-based) in a gardening or horticultural context.

So, for the purpose of this book, my own definition of compost (and composting) is: a mixture of dead, mainly plant material from the garden, which is undergoing a process of decay. This requires a range of physical, chemical and biological processes involving a variety of microorganisms (bacteria and fungi) and small invertebrate animals. The nature of the compost will depend on the stage of decomposition as well as the composition of the starting material.

3 / WHY COMPOST?

This question has been touched on already but let's go further. It is spring in the garden, and the enthusiastic gardener has taken a first cut of rather long grass from the lawn. Or maybe it's autumn: the leaves are tumbling from deciduous trees and there is a mass of cutting back to do of plants that have finished for the summer. What do you do with all that organic 'waste'? In many areas there are kerbside collections of so-called green garden waste. However, even if this sort of disposal is an option, you might have the uncomfortable feeling that using it is not quite the right thing to do. Recycling comes to mind – what was that lesson in school biology or geography about cycling of nutrients? And how often do you encounter the word recycling at the supermarket and elsewhere? So, if nothing else, a compost heap or bin will improve your environmental credentials. If you get it right, not only will the volume of green waste reduce substantially but the composted residue will be beneficial to return to your flower beds, vegetable plots or flowerpots. That material should do two things. First, it will put back some of the nutrients, such as the element nitrogen, that growing plants have taken out of the soil during the summer. Second, the addition of organic matter should also improve what is referred to as soil structure. Essentially this means changing a heavier soil into a lighter one – so a soil with a more open texture, with more pores and air spaces. This can be as beneficial to plants as the addition of extra nutrients. For healthy root growth and a well-grown plant, the soil needs a steady supply of oxygen from above ground: a lighter soil is more permeable to oxygen from above ground. So, composting is a good idea. It is ecologically and environmentally a positive action, is good for your garden plants – and may even save you some money on buying commercial compost.

COMPOSTING IS DECOMPOSITION – THE ROTTING OF ORGANIC MATTER

4 / WHAT CONTROLS THE ROT?

The key process in composting is decomposition - the rotting of organic matter. This is a complex combination of energy transfer and the release and cycling of nutrients. But what are the factors that control this process? This question is central to effective composting and will be explored in the coming sections. It turns out to be an involved and nuanced story, but if you were to try to model what goes on, you would require three strands of information about the compost system: 1) the physicochemical environment (P) - factors such as temperature, acidity and moisture; 2) the quality of the organic waste resource (Q) - nutrient content and its accessibility; 3) the decomposer organisms/community (O) - soil invertebrates and microorganisms. These components and their interactions are illustrated in Figure 2. The various permutations of these factors dictate how quickly or slowly mixed compost material will decompose. Decomposition can be briefly summarized as the partial conversion of organic (that is, biologically created) matter to (mostly) soluble inorganic molecules and ions. This latter process is referred to as mineralization and makes nutrient elements available to plants while leaving 'difficult' undecomposed material in the mature compost.

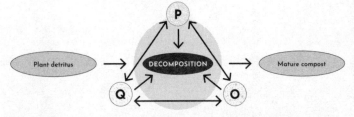

Figure 2. Factors that control decomposition processes in compost: P = physicochemical factors; Q = nutrient content and accessibility; O = decomposer organisms. Each factor has an individual effect, but rates of decay will depend on the combined effect of all three.

A VARIETY OF PLANT ORGANIC MATERIALS 'FEED' THE COMPOST HEAP

5 / DISSECTING A COMPOST HEAP: ORGANIC MATTER

In the garden, waste plant organic material has been collected and is now in a pile or a bespoke compost bin. For the present, let's assume that there is little or no kitchen waste in the bin, just a collection of dead plant material. It is the product of lawn cutting, leaf raking and dead plant clearance. What are the components of this plant waste?

A variety of plant organic materials 'feed' the compost heap. In most gardens, a good part of it will be fallen tree leaves (or leaf litter, as it is known). As previously mentioned, in temperate regions there is the regular change of season in the autumn; this promotes the change of colour of deciduous trees, which is followed by leaf fall. In warmer climates, other patterns of leaf fall apply but the leaves still fall.

Many trees and shrubs shed their leaves rather than trying to sustain leaf cover through the winter. The changing colour of leaves in the autumn signals that the leaves are dying as part of this process – the biological term for which is senescing. However, a lot of resources have been channelled into growing foliage in spring and summer and this necessitates a recycling process within the tree, to recapture materials that would otherwise be lost. For example, the protein known by the acronym RuBisCO (or Rubisco) is the most abundant protein in leaves: it can represent half or more of the total leaf protein. It is also the most important catalyst (enzyme) in the process of photosynthesis. During senescence, RuBisCO is broken down by protein-digesting (protease) enzymes to allow the molecule's nitrogen content to be retained.

So, the autumnal change of leaf colour confirms the process of withdrawing useful nutrients such as the nitrogen derived from RuBisCO from the tree leaves. By the time the leaves tumble from the tree, they represent a rather asset-stripped resource. However, all is by no means lost for decomposers. The extent to which nutrients are moved (translocated) out of the leaf and the efficiency of their recovery (resorption) varies with conditions and plant species. A review of the scientific literature suggests that there is an average of 50 per cent resorption of nitrogen and 52 per cent of phosphorus (Aerts, 1996). Another study showed that resorption is also high for potassium, but much less so for calcium and magnesium (Hagen-Thorn et al., 2006). In other words, the degree to which different elements are recovered also varies. Generally, there seems to be a strong correlation between nutrient concentrations in green leaves and the amount of nutrient that remains at leaf fall. For example, higher green-leaf levels of nitrogen will also be evident in the senesced leaves (Cornwell et al., 2008). But even after all this translocation activity, the basic structural 'scaffolding' of the leaf remains unchanged.

The point about the movement of nutrients is true for perennial deciduous plants (trees and shrubs), but the situation with shorter-lived plants (annuals, biennials or herbaceous perennials) is rather different. Generally, they do not have permanent woody tissue above or below ground in which they can store withdrawn nutrients. So, to a large extent, these nutrients remain in the leaf tissue as it dies. Nevertheless, the browning or yellowing of those leaves at the end of the growing season shows there is a kind of biochemical 'anarchy' comprising a breakdown of organization in the leaf cells and molecules. This is likely to happen before the leaves get to the compost and so will change the nutritional content of the leaves. The exception to this pattern relates to perennial plants with persistent roots, bulbs or corms – each of which is a storage organ. In that case, there will be transfer of nutrients from the aerial parts of the plant to underground. This winter store of nutrients then provides for rapid growth the following year. So, the senescent leaves of plants with persistent (perennating) storage organs are also likely to be nutrient deficient, as with trees and shrubs.

So, what is in the 'structural scaffolding' remaining in the dead leaves, once translocation has been completed? Plant litter contains six main categories of chemical constituents: 1) cellulose; 2) hemicellulose and pectin; 3) phenolics (lignin and tannin); 4) water-soluble sugars, amino acids and other small organic molecules; 5) lipids (fats), oils, waxes, resins and pigments; 6) proteins (after Satchell, 1974).

The following sections provide a brief explanation of each of the categories listed, except for the smaller water-soluble molecules (4), which are relatively familiar.

Cellulose

This is the single most abundant and ubiquitous plant-construction molecule. To describe it in more familiar dietary terminology, cellulose is a large carbohydrate molecule. Chemically this means that its molecular structure contains only three types of atoms/elements – namely carbon, hydrogen and oxygen. Cellulose is the key structural molecule found in the cell walls of plants (and algae). As an analogy, you can think of a plant cell as a box with strong walls, and a balloon filled with water squashed inside the box. Cellulose forms the walls of leaf cells (the box in our analogy), inside which sit water-based cell contents (called cytoplasm) within a cell membrane (the balloon). This is true of both the softer leaf cells and tissues as well as the cells forming the leaf veins, although these are constructed with tougher cell walls that contain more than just cellulose. Cellulose can also be described as a complex biological polymer. A polymer is defined as being constructed from (fairly) small, relatively simple molecules that are then strung together in tens or hundreds or thousands (like a very long necklace!) to create a much larger molecule. In the case of cellulose, these long polymeric molecules are bundled together into fibrils, which, laid in parallel, represent the main construction material in plant cell walls (see Figure 3).

But there is a particular paradox with cellulose. The starting point for making cellulose polymer is a form of the familiar glucose molecule. Glucose is well known as a sugar that is easily consumed and digested. It is typically touted as a quick intake of fuel for our bodies, as in so-called energy drinks. However, once assembled into a long-chain cellulose molecule (also known as a polysaccharide molecule – which is to say, many sugars), it can no longer be easily digested. So, for us, it has no nutritional value, aside from being a useful source of dietary fibre. Fortunately, other types of organisms involved in decomposition can break down cellulose and use it as an energy source – about which more later.

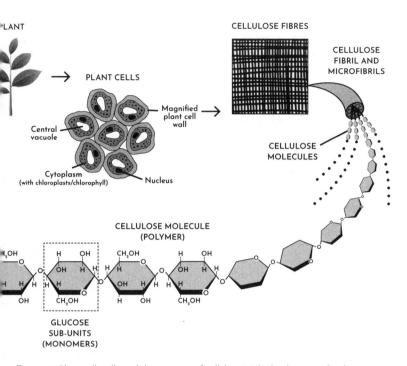

Figure 3. Plant cell walls and the structure of cellulose. Multiple glucose molecules link together to form a polymer polysaccharide (= many sugars), which are then bundled together into microfibrils, which in turn are bundled into fibrils and then into fibres, which create the basic 'scaffolding' structure of plant cell walls.

Hemicelluloses and pectins

Also present in plant cell walls is a related family of short-chain polymers (also polysaccharides) known as hemicelluloses. These differ from cellulose in being constructed from sugar molecules other than glucose, and they provide infill material between cellulose fibrils. Their texture is referred to as amorphous – rather soft and with no clear structure – which is also different to that of cellulose.

Pectins form a further complex class of polysaccharides present in plant cell walls. Their well-known role as setting agents in jam making hints at their gel-like characteristic. Beyond the plant cell walls, pectins also fill the gaps (middle lamella) between adjacent cells, serving as a kind of glue to hold plant cells together.

Lignins and tannins

Despite the single generic name for this material, lignin is a family of large, branched polymers that vary in structure between plant species and can be fiendishly complex. Like cellulose, these molecules comprise only carbon, hydrogen and oxygen atoms. However, their particular chemical

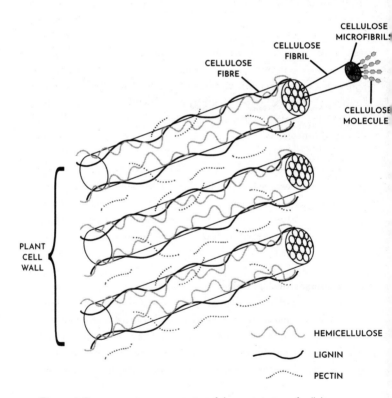

Figure 4. Diagrammatic representation of the organization of cellulose, hemicellulose and lignin in plant cell walls – known collectively as lignocellulose.

structure makes a tough material that confers strength to plant tissues when inserted into cell walls between the cellulose and hemicellulose molecules (see Figure 4). It is also rather resistant to microbial decay. Leaves are not the best place to look for lignin, but it is abundant in supporting tissue and especially in any woody tissue that finds its way into the compost heap. Lignin represents 15-35 per cent of the wood of most trees (Satchell, 1974) – and wood is of course recognized as a tough material; indeed, it is wood that enables trees to stand at 100m/ 330 ft or more in height. At the same time, were you to examine a cross-section of a non-woody plant stem you would likely find it includes a tissue type known as sclerenchyma – the cell walls of which are similarly strengthened with lignin. In other words, most plants would be very floppy without the strengthening presence of at least some lignin.

Cellulose, hemicellulose and lignin are collectively referred to as lignocellulose and are the major components of plant cell walls. The take-home message here is that plant cell walls can contain a mix of materials, some decidedly tough and others rather soft. Their fate will be followed as decomposers get to work on them in the compost heap.

Tannins are another class of large, polymeric molecules that some plants (for instance, tree species like oak) are especially rich in – although tannins do not have structural functions. You might associate the name with the word tanning: tannins were originally extracted from oak bark and used to tan leather – a characteristic relevant to their fate in decomposition. Tanning is a way of stabilizing fresh animal skins by creating 'bridges' between the skin collagen proteins, which helps turns it into tough durable leather. If you measure tannin levels in oak leaves, they gradually increase during the growing season to the extent that by the end of the summer they may represent 50 per cent of leaf mass. They are mostly soluble and so can be washed out (leached) from leaf litter lying on the ground. We will come back to tannins later to look at how they themselves decompose and influence the decomposer community.

So far we have looked at only carbohydrate (polysaccharide) molecules in plant detritus. Prevalent though they are in dead leaves, they are not the only molecules – and carbon, hydrogen and oxygen are not the only elements in dead leaves. If they were, they would provide a rather poor diet for any organism wishing to live off leaf litter.

There isn't room here to consider the full scope of molecules present in living leaf material. However, a very brief review is sufficient to give a broad idea of the elemental composition of other major molecules and hence their potential availability as substrates for decomposition.

Lipids

These molecules are also structured with carbon, hydrogen and oxygen atoms forming single, paired or triple chains. The triple-chain lipids are referred to as fats. Another essential category of lipids – phospholipids – are a major component of cell membranes and contain the element phosphorus (phosphate ion – PO_4^{3-}). Several plant lipids serve an important function as 'energy storage' molecules, such as is the case in seeds with a high oil content (for example, sunflower and rapeseed).

Proteins

These are formed as linear chains of amino acid sub-units. As well as carbon, hydrogen and oxygen, a key feature of proteins is the presence of a nitrogen-containing amino group ($-NH_2$). Other elements such as sulphur may also be present in specific proteins. Proteins can be structural or have a functional role in metabolism, such as is the case with enzymes.

Nucleic acids

Nucleic acids (DNA, RNA) are large complex molecules which are a store of genetic information in cells. As well as a core structure made up of carbon, hydrogen and oxygen, there are four different nitrogen containing sub-units (nucleobases) and a phosphate group, which contains phosphorus.

When it comes to thinking about which of these chemical components might remain in dead leaves, things become a bit more difficult because there is no single answer. As leaves die, the fate of the remaining cell contents becomes key. The living part of the plant cell (as opposed to the fairly inert cell wall) contains the mix of molecules described above along with various mineral ions and other molecules such as vitamins, pigments and enzymes, which have a variety of roles. Many of these types of molecules will be common to all plants but individual plants species have their own biochemical and physiological complexities and adaptations requiring a suite of additional molecules that may be present in some of their cells. The most important point to note at this stage is that collectively these various molecules will contain some very important chemical elements incorporated in their structure. Some of these elements are present in rather large quantities (especially nitrogen, phosphorus, potassium, calcium - known as macronutrients), while others are available and needed only in small amounts (for example, iron, magnesium, molybdenum - referred to as micronutrients). All in all, a 'typical' plant is likely to be made up of more than twenty elements - which should be good news for any potential decomposer organisms in our compost heap. But, as already noted, dying tree leaves are asset-stripped by the parent plant before falling to the ground. Although not everything is removed, many of the elements discussed so far will now be present in rather low concentrations. This includes those elements that are needed in relatively large amounts - such as nitrogen. So exploitation of nitrogen and other elements present in dead leaves can be key to effective decomposition.

We have focused on dead leaves as a major component of a compost heap, but it is likely there will be other herbaceous material and plant roots too. There could also be lignin-rich material, especially twigs and stems pruned from woody plant species, as well as rather less lignin-packed grass clippings and so on. Unlike dead leaves, grass will be cut fresh and tends to be added to the compost heap soon after cutting; for this reason, fresh grass is a much more promising substrate for decomposition. After all, you can make a cow out of grass! The rich mix of molecules and nutrient elements found in living grass will still be present as the grass is tipped on to the compost heap.

What else will be added to a compost heap from garden refuse? Aside from organic plant matter, at very least the compost heap will include lesser or greater amounts of soil, typically attached to bedding plant roots or discarded potted plants. That soil will include a large percentage of non-organic mineral material (such as larger particles of sand and minute particles of clay), but also decayed plant material and perhaps peat or other commercial compost used in growing or potting. There may even be artificial but rock-derived materials like perlite or vermiculite, which are used as additives to potting compost. As such, the mineral matter will contribute to the feel or texture of the final composted material.

This is still not quite the end of the story, though – because the material added will be seeding the compost heap with a variety of living organisms, most but not all of them small and largely invisible. And it is these living organisms that power the decomposition process. As we now know, from the outside the compost heap will seem static, but this belies the microscale activities and transformations that a cursory probe of the compost starts to reveal. As perfectly described in a *New Scientist* article title, compost is 'a ferment of little rotters' (Anderson, 1983). It is these little rotters that we will consider next.

AN ARRAY OF LIFE IN THE HIDDEN WORLD OF SOIL AND COMPOST

6 / DISSECTING A COMPOST HEAP: LIVING ORGANISMS

Above ground, some compost organisms would be easy to spot – larger invertebrates, for instance, and (to some extent) microorganisms such as fungi. But the subterranean world of a compost heap reveals its living organisms more guardedly: they must be searched for. At one time the term 'cryptosphere' was used to describe this hidden underground world of soil and compost – where the soil animals are classed as 'cryptozoa' – but the name has now been taken over by proponents of cryptocurrencies!

It is increasingly accepted that soils and compost make a substantial contribution to world biodiversity. This chapter is about sorting out the living characters in the hidden world of soil and compost before we go on in later sections to consider the roles they all play in the composting process. We can begin with the microbial component, which is both the most diverse and, in many ways, the least known. There is such variety in the microbial world that it can readily offer up new discoveries to researchers. A fascinating recent example is the discovery of a deep subterranean microbial world powered not by the sun, but by radioactivity (Cepelewicz, 2021). The microbial world in soil and compost might be better known, but even here the astonishing diversity encountered in a few grams of soil or compost represents a major scientific challenge. This diversity will vary with soil and compost type, geographical location and climate; there is no universal suite of microorganisms to be found across terrestrial environments (Bardgett and van der Putten, 2014). A study by Fierer and Jackson (2006) across a range of locations in North and South America found soil pH (rather than geography) as a major factor on bacterial diversity. So local conditions in compost might have a greater effect on bacteria than location per se.

Microbial passengers

As already discussed, we can assume that dead leaves and cut grass will generally be the largest component of a garden compost heap. Although it is not immediately obvious, the surfaces of plant leaves harbour a range of fungi and bacteria that can be isolated from leaves almost as soon as they unfurl. Dead leaves therefore can and do bring fungal and bacterial passengers with them to a compost pile. These include fungal species that later, after the leaf dies, contribute to leaf decomposition; they start by growing just on the leaf surface and then, as the leaf senesces, they invade its tissues (Voříšková and Baldrian, 2013). Some fungi possess enzymes (such as chitinase) which help them penetrate the outer waxy cuticle layer of a leaf. When on the leaf surface, many fungi take the form of fine, essentially invisible living 'threads' (or hyphae) made up of strings of linear cells. These represent the typical life form of a fungus but become visible only when the threads grow densely together – then you can see it as grey mould, which can appear on bread, or as the blue-green fungal felty layer on oranges. Some types of fungi, the yeasts, have a different growth form: as single oval cells that reproduce by budding off new cells. Yeasts can also be found on the leaf surface and, along with the hyphal fungi and leaf surface bacteria, are collectively known as 'phylloplane' microorganisms. Together they form a small microbial community, a leaf microbiome. In those places and habitats where leaves grow in rather damp and humid conditions, the phylloplane may also play host to algae and lichens. The latter organisms are a symbiotic mix of algae, fungi and bacteria and will typically form a visible 'crust' where they occur on the leaf surface. Unlike fungi and bacteria, neither algae nor lichens will persist within compost because they have a requirement for light.

A very different microbial community is present on the roots of plants added to the compost heap. Although very much part of the soil environment, root surfaces and the surrounding soil (rhizosphere) harbour a particularly rich microbial community (root microbiome) which is greatly influenced by molecules exuding from the plant roots. Up to one-third of carbon fixed by plants in photosynthesis (as sugar) may be released from plant roots into the soil and provide a food source for the

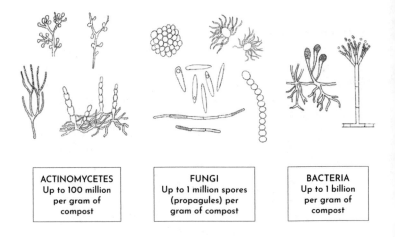

| ACTINOMYCETES | FUNGI | BACTERIA |
| Up to 100 million per gram of compost | Up to 1 million spores (propagules) per gram of compost | Up to 1 billion per gram of compost |

Figure 5. Major groups of compost microorganisms. The abundance data should be taken as a broad indication of numbers and relative abundance rather than as definitive. There will be wide variations between composts and variations in laboratory methods of counting. There is more discussion on this topic in the text. Data from Miyashita et al. (1982), Horwath et al. (1995), Whitman et al. (1998) and Wang et al. (2007).

root microbiome (Heal et al., 1997). Although not directly relevant to composting, this root exudation will affect the kinds of microorganisms attached to roots.

Endophytes

Microorganisms on plants are not restricted to outer surfaces. In recent years it has become clear that there is an extensive suite of fungi, referred to an endophytes, that can colonize internal plant tissues of both herbaceous and woody species. Although many of these may be benign and appear to do little harm to their host plants, there are others for which the boundary between cohabitation and parasitism is less clear cut. Many of the fungal species are also known to be potential saprophytes (living off dead organic matter) or pathogens (disease causing). So, there will be senescent plants infected with their endophytic or pathogenic organisms which may then contribute to the early stages of leaf litter decay (Voříšková and Baldrian, 2013). A study of beech (*Fagus sylvatica*) litter showed that many of the endophytic fungi present in green leaves continued to be active in one-year-old beech litter and were connected to the process of decomposition (Guerreiro et al., 2018). On the other hand, there will be other endophytic fungi that will be lost in the transition to a compost environment.

Fungi in wood

Much of the discussion about senescing leaves and their microorganisms will largely apply to freshly cut grass – with the exception of lichens, which are not normal inhabitants of grasses. But the story is different with the woody material swept up from the garden or pruned from trees and shrubs. This may already be infected with a range of wood-rotting fungi. A pink coral-spot fungus (*Nectria*) is frequently seen on dead branches and twigs before they fall from the tree or shrub. The visible pink spots are the fungal sporing structures, while the bulk of the fungus will be buried beneath the bark. So, if you peel back the surface bark from a woody branch that has been dead for some time, then the tell-tale sign of fungal hyphal strands may be visible, especially in damp wood. Sheets of such hyphae are known collectively as mycelium. The coral spot is typically a saprophyte, which means it is living off the dead branch tissue. These are the types of fungi that are the most likely to be encountered in a compost heap, rather than those causing disease (pathogens). However, even *Nectria* species are not averse to invading living tissue if they encounter a susceptible host plant. So, the boundaries for some fungal species between endophyte versus pathogen versus saprophyte are not necessarily well defined.

How many microorganisms?

And what of the soil that inevitably finds its way into the compost heap? This could be carrying a rich assemblage of life. There will be a lot of fungi and bacteria – which both overlap and are complementary in their decomposition activity. In terms of abundance, some amazingly high values are quoted for numbers of bacterial cells that each gram of soil may contain. Different soils and different methodologies can produce different figures, but 100 million bacterial cells per gram of soil would not be unusual, with some published figures much higher (Whitman et al., 1998). Limited data from compost generate similar figures (Horwath et al., 1995). Values for fungi are typically quoted in terms of metres of hyphae per gram of soil. On that basis, numbers range from 3 to 30 m/10 to 100 ft (Shen et al., 2016). Alternatively, Horwath et al. (1995) quoted 1-10 million fungal propagules (typically spores) per gram of laboratory composted grass straw. If you think about it, both the bacterial counts and fungal hyphal length are really quite astonishing. A gram of soil will vary in the volume it occupies, depending on the nature of the soil. But a typical value of 1 g/0.04 oz soil = 1 cm^3/0.06 in^3 would not be unreasonable. Imagine trying to fit 100 million bacterial cells or 30 m/100 ft in length of fungal hyphae into that volume!

Apart from 'normal' bacteria, there are other groups of microorganisms present. For example, the so-called filamentous bacteria or actinomycetes, known as a source of antibiotics. They are also thought to be the cause of the earthy smell of soil and compost and contribute to the decomposition of 'difficult' organic materials like lignin. Their filamentous pattern of growth can appear similar to that of fungi. Lastly, photosynthetic blue-green bacteria (cyanobacteria) may occur if the soil comes from the surface and has been exposed to light.

A note on species

Numbers are one thing, but what about microbial species? The addition of soil to compost is likely to introduce a wide spectrum of microorganisms and is more likely to include those species adapted to decomposition processes than those arriving on leaf surfaces. Estimates of the number of bacterial species in soil have proved problematic, but the indications from RNA/DNA analyses are that there can be several thousand bacterial types in just one gram of soil (Schloss and Handelsman, 2006). The application of RNA/DNA analysis techniques has made it increasingly difficult to define the boundaries of bacterial species, leading to the increased use of the term Operational Taxonomic Unit (OTU), avoiding the use of species terminology.

The variety of small soil invertebrate animals referred to collectively as soil fauna may also arrive in any transferred soil – though some are barely visible. If you think of pond dipping as an exercise in exploring the hidden depths of the aquatic world, then a sample of soil – with attached organisms – is a bit like soil dipping. We will come back to the details of what might be encountered later. Here, a brief listing provides a sense of the key groups that may be hitching a ride in soil added to the compost pile. It is quite likely to include mites, springtails, slugs and snails, earthworms, woodlice, millipedes and centipedes and other types, generally less common and/or less visible. If the compost heap or bin is placed directly on soil, then larger numbers of invertebrates can gain direct access from below to any fresh input of grass or leaf litter. It is worth noting too that some vertebrates – such as hedgehogs, mice, newts and snakes – may make a home in a compost heap, which will be an especially cosy hiding place over the colder months. This has little impact on the business of composting but is something to be aware of when you turn your compost heap over, especially during winter and spring.

7 / MICROORGANISMS AND DECOMPOSITION

Many and varied processes contribute to decomposition. What exactly happens to turn dead leaves into soil-conditioning compost? The leaves and other organic matter are quickly colonized by an array of microorganisms (microbiota) that can gain all the nutrition they need from this rather depleted food source. This mode of feeding classes these organisms as saprophytes, and the mode of nutrition as saprophytic. The microbes are the key drivers of nutrient cycling in compost. Given time, the variety of microbial species present in compost and their diverse biochemistry have the collective ability to decompose almost any organic material they are challenged with (Satchell, 1974). Recently the extent of microbial decomposition adaptability caused excitement with the discovery of natural microbial enzymes (called cutinases) that can break down PET plastic used in the manufacture of plastic bottles. A particularly promising enzyme is even known as a leaf-branch compost cutinase (LCC) (Tournier et al., 2020). At the same time it has also been suggested that even 'compostable' bags based on starch often do not decompose readily in garden compost bins (Vaverková et al., 2014) – although it would be interesting to test their decomposition in a garden bin for longer than the twelve weeks of that study. In any case, I wouldn't start adding plastic bottles to compost heaps just yet.

Microbial succession

To begin with, some readjustment of thinking about compost is required, from the descriptive and rather static to the active and dynamic. Without drifting too far into cliché, think of a changing kaleidoscope of microorganisms and invertebrates that migrate, colonize and die in the organic matter as decomposition progresses. To use the technical term, there will be a succession of organisms that are adapted to different stages of the decomposition process. The first arrivals will be fungal and bacterial species, which will mop up the relatively nutrient-rich remains of the cell contents. The general term for the main part of the cell contents is cytoplasm, which in the living cell will be a mixture of proteins, carbohydrates and lipids and a variety of other molecules. Apart from containing a cell nucleus, the living cell cytoplasm also houses a range of so-called organelles. These include the mitochondria, referred to as the powerhouses of the cell, and chloroplasts, where the light-capturing pigments (chlorophyll and others) are lodged. In the middle of the cell is a membrane-bound vacuole, with a water-based liquid interior referred to as cell sap. This can occupy a large part of the volume of a plant cell. Going back to our earlier balloon-and-box analogy, the vacuole will be like a second balloon inside the first: the outer balloon represents the membrane containing the semi-solid cytoplasm, and the inner balloon separates the liquid-filled vacuole from the cytoplasm. The vacuole is largely a storage organ for water and various materials that are kept away from the cytoplasm. However, as plants senesce and cells die, the cell organelles are progressively dismantled – starting with the chloroplasts. Other organelles are still needed to transport nutrients out of the cell, so these structures survive longer. At the end of the resorption process, remnants of the cytoplasm are still present and are a nutritional prize for microorganisms that colonize quickly and have a fast growth and reproduction cycle. In due course, they are replaced by other, 'slower' microbes.

Figure 6 (see page 41) demonstrates the concept of succession on dead leaf material, with particular reference to fungi. Although the names may be unfamiliar, one of the 'early' fungi is *Mucor*, which grows fast and produces spores (sporulates) quickly; a very similar fungus, *Rhizopus*, can be encountered in the home as black bread mould. If you keep a piece of bread in an enclosed, moist environment, there is a fair chance that the grey felty mycelium of *Rhizopus* will appear. Look closely with a magnifying lens (or maybe a magnifying app on your phone) and you should see tiny black structures, which are spherical spore-bearing sporangia. Also in Figure 6, this time at a late stage of decay, you will note a fungus genus called *Penicillium*. As the name suggests, the antibiotic Penicillin was first isolated from a member of this genus. You are also likely familiar with it as the typical cause of decay in citrus fruit. Once again using a magnifying lens or microscope, a look at the felty blue-green layer on a rotting orange will confirm the presence of coloured fungal hyphae. These grow into brush-like structures bearing clusters of tiny spores, which will be released in hundreds of thousands and disperse to new substrates to colonize – perhaps to another orange in the dish. In relation to decay succession, this genus of fungus appears later in the decay sequence but will similarly sporulate in the compost on a suitable substrate. The spores may stay local within the heap but, if you turn over the compost regularly, hundreds of thousands of them will be spread by air currents into the external environment. Even within the heap, both fungal spores and bacteria are likely to be carried by invertebrates such as earthworms to freshly added organic material. But fungal spores and bacteria are ubiquitous, so it is unclear to what extent dispersal by soil invertebrates is required or significant for colonizing fresh material. It may help.

All this prompts the question of why some fungi arrive late to the decomposition party. Of course, the simple reason might be that some fungal spores disperse more easily than others or are maybe more abundant in the environment. Perhaps there are differences in the competitive ability of different fungi. Either could explain the quick appearance of *Mucor*. However, after this fungus has consumed the easy material, it then gets pushed out by more specialist fungi – which is to say, fungi that are able to decay more difficult (recalcitrant) plant material. We have already noted some of these tough polymers in the plant cell walls. If they can be broken down, they can provide a source of energy as well as the carbon required to grow new fungal and bacterial cells. So, a quick reminder that cellulose is a polymer carbohydrate (polysaccharide). Its decomposition relies on fungi having the right armoury of biological catalysts (extracellular enzymes) released from the cells to depolymerize this molecule. Enzymes are specialist molecules that have remarkable abilities to promote the fracture of specific types of complex molecules. So-called biological washing powders containing enzymes use this trick to remove a variety of organic stains from clothes.

All biological organisms rely on enzymes of many kinds. When eating, the first thing that happens in our mouths is the mixing of the food with saliva, which contains a starch-degrading enzyme called amylase. This is the same enzyme used by microorganisms to decompose starch carbohydrate in organic waste - maybe in an old potato thrown onto the compost heap. But cell wall material is much tougher than starch. Cellulose requires a cellulase enzyme to split the molecule, while lignin requires a lignase. (This can also be referred to as a ligninase and the current label of choice is lignin-modifying enzyme (LME). We will stick to lignase for ease of use.) Note the ending 'ase', which is used to denote an enzyme molecule. Each of these terms covers a class of enzyme rather than just a single type of enzyme molecule. Cellulase and lignase are not typically synthesized by animals (but more about this later), so this aspect of decay is very reliant on fungi and bacteria. However, there are other enzymes that fungi and bacteria secrete - such as proteases to digest protein and lipases to digest lipids - that can also be produced by animal species, including those involved in decomposition.

Wood decay

We have so far focused on the decomposition of leafy and herbaceous plant material but, as previously noted, compost is also likely to include some woody material. Decomposition of wood typically involves an additional suite of fungi, especially those belonging to a group (taxon) known as Basidiomycetes (Basidiomycota). This group includes the familiar mushrooms and toadstools, which are the spore-producing structures of a fungus growing much more extensively as mycelium in the soil and litter. The reproductive role of the mushroom can be demonstrated by placing the cap of a mature mushroom or toadstool overnight on a sheet of white paper. In the morning you will find a spore print that traces out the pattern of the gills on the underside of the cap (some mushrooms have pores instead). The spores have been shed from the gills and dropped on to the paper. Where there are air currents, the spores will be dispersed widely rather than falling on to home ground.

The mushrooms or toadstools that can be seen above ground are just the tip of the iceberg, so to speak - the bulk of the fungal mycelium is found threading through the soil and litter layers. As this mycelium expands, it can encounter fresh woody material to colonize. This is a kind of search-and-destroy mission. Once the advancing fungal hyphae encounter a suitable substrate, further hyphal growth is diverted to this zone. But the role of fungi in decomposition varies with species. In general, the wood-rotting fungi can be classed as either 'white-rot' or 'brown-rot' species. The former tend to be found in deciduous wood and are adept at degrading lignin, giving the wood a bleached appearance. The brown-rot fungi are more typically found on coniferous wood and mainly decompose cellulose and hemicellulose. Rather than using enzymes, brown-rot fungi release and utilize a strong oxidizing compound, hydrogen peroxide, to break down the cellulose/ hemicellulose. Because they leave behind much of the lignin, the wood retains a brown colour. An example of brown rot is the so-called dry-rot fungus *Serpula lacrymans*, which can cause extensive rotting of house timbers. More apparent and possibly more familiar is a white-rot fungus called Turkey Tail (*Trametes (= Coriolus) versicolor*). This multi-layered bracket fungus, patterned with concentric rings, is frequently found on fallen tree trunks in woodlands or gardens. Another well-known white-rot fungus is the edible oyster mushroom (*Pleurotus ostreatus*), now extensively cultivated and widely sold in supermarkets. Wood on the surface of a compost heap might similarly show mushroom or toadstool fruiting bodies. These and other fungi contribute to a sequence of fungal species colonizing and decomposing wood, as was the case with herbaceous/leaf material. A reminder that this is the ecological process of succession.

So, there is an explanation for why fungal decomposer succession occurs. Horses for courses – or, to put it another way, specific fungi for specific decomposition activity: colonizing in sequence.

Bacteria

You may have noticed that so far there has been little mention of bacteria. This is mainly because of the complexity of pinning down the activity of so many different types of bacteria to so many different roles. However, it is well known that bacteria can produce a wide range

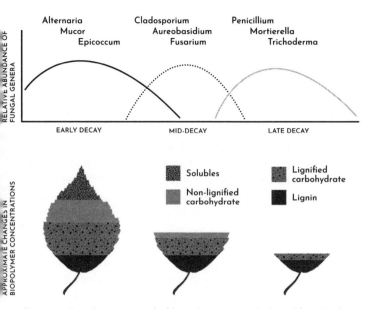

Figure 6. Fungal succession in leaf litter decomposition (adapted from Vivelo and Bhatnagar, 2019).

of enzymes necessary for decomposition. Remember the earlier point about grass being turned into a cow? This means the cow must break down the cellulose-rich grass as efficiently as possible – and yet animals do not generally produce cellulase enzymes. This dilemma is solved by the design of the cow stomach, which is like a huge fermentation vessel in which a rich 'soup' of many types of bacteria digest the grass – especially the copious quantities of cellulose. This digestion strategy is true of herbivores generally – they need their symbiotic gut bacteria to make good use of their vegan food. Similarly, some bacteria in a compost heap can enzymically break down cellulose, albeit not as efficiently as fungi. They are primarily active in the degradation of simpler polysaccharides. But there is still a succession of bacteria on fresh plant detritus, with specialist bacteria serving the different stages of decomposition. Yet this is not quite the end of the process of decomposition. Bacterial and fungal cells themselves become substrates for further decomposition after death – by other bacteria and fungi (López-Mondéjar et al., 2018).

Antibiotics

A discussion of soil and compost bacteria also leads on to the subject of antibiotics. These play a key role in human health and have been isolated from a variety of soil microorganisms. For example, hundreds of different molecules with antibiotic properties have been isolated from a genus of filamentous soil bacteria (actinomycetes) called *Streptomyces*. Although the production of such antibiotics can be demonstrated very easily in pure culture in the laboratory, showing that they have a role in microbial ecology in soils has proved more troublesome. If you have a vision of 'microbial wars' in the compost heap, with antibiotics as ammunition, there is little research evidence for these kinds of interactions. Any natural levels of antibiotic substances are in low concentration, which has led some scientists to conclude that the interpretation of them having a defensive function is wrong. Rather than providing a competitive edge against other microorganisms, one suggestion is that these molecules have a key role in cell-to-cell communication (Davies, 2006). Davies argues that the function of these small molecules is for microbes to 'talk' to each other. But quite what is going on isn't clear.

Figure 7 (right) A soil invertebrate community in an area of old grassland in Finland (Törmälä, 1979). The abundance of invertebrates in a compost heap is very variable but will broadly reflect the major differences in numbers and biomass shown here. A mathematical transformation (conversion to logarithmic values) of the data has been used in figures 7(b) and 7(d) to clarify the relative abundance of the different groups. The actual number of individuals per square metre reported for a July sampling were as follows: roundworms – 11,260,000; potworms – 47,000; earthworms – 300; spiders – 307; mites – 53,710; springtails – 28,590; beetles (adults + larvae) – 703; fly larvae – 1,553. Other invertebrate types may be present in compost in smaller numbers but are not included here.

A. NUMBER PER SQUARE METRE

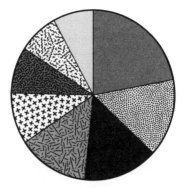

B. NUMBER PER SQUARE METRE (LOG SCALE)

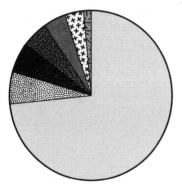

C. DRY WEIGHT (BIOMASS) MILLIGRAMS (mg) PER SQUARE METRE

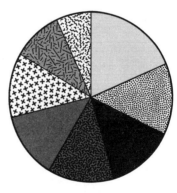

D. DRY WEIGHT (BIOMASS) MILLIGRAMS (mg) PER SQUARE METRE (LOG SCALE)

 Earthworms

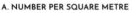 Mites

 Spiders

Roundworms

Springtails

Beetles - adults + larvae

Potworms

 Fly larvae

Although soil microorganisms are considered to deliver the bulk of leaf litter decomposition in a compost heap, there is also the soil invertebrate fauna to consider. Do they all 'do' decomposition? Or do we need a more nuanced question? As already suggested, there will be a wide variety of invertebrate types in compost - presumed to be doing different things. Invertebrates therefore need to be discussed by individual type rather than collectively.

MICROBES ARE KEY DRIVERS OF NUTRIENT CYCLING IN COMPOST

8 / INVERTEBRATES AND DECOMPOSITION

How many invertebrates?

Counting invertebrates in compost is one thing - labelling them with species names is more problematic. The challenge is not as great as with fungi and bacteria, but on a global scale there will be many groups, such as roundworms and mites, where knowledge of species is very incomplete and expertise in the taxonomy of these groups can be hard to find. Although there is much overlap of invertebrate types found in soil (and compost) across the globe, some have geographical bias. One clear example of that is termites, which are missing from temperate biomes. This section focuses on numbers and relative abundance of broader groups (taxa) rather than species. Can we make any generalizations about the invertebrate types present in compost and their relative importance to the ecology of the compost ecosystem and decay?

Figure 7 (see page 43) represents data for the kinds of invertebrates that are found in European grassland leaf litter and soil. This is a snapshot of invertebrate abundance in one particular soil type - rather than in compost, for which there are few quantitative data on the full range of compost invertebrates. Whether soil or compost, numbers of invertebrates will generally be very variable. Bearing this in mind, the numbers in Figure 7 should not be taken too literally. But they do tell us something about the relative abundance of different invertebrate types. More recent data from Bardgett and van der Putten (2014) provide a more synoptic view of soil invertebrate numbers from a range of publications (expressed per square metre of soil/litter): roundworms - 200,000-9,000,000; potworms - 12,000-311,000; earthworms - 300; millipedes - 110; woodlice - 10; springtails - 10,000-50,000; mites - 10,000-100,000. With either data set, it is clear that roundworms (nematodes) dominate in numerical terms: there can be several million in each square metre of soil. But change the data into biomass and a rather different picture emerges – earthworms are now the major group. In fact, Figure 7 shows roundworms as only fifth in the sequence of biomass abundance. This is because soil/compost invertebrates span a very wide range of sizes – from the very visible (such as earthworms) to the microscopically small. In terms of body length (see Figure 8), the tiniest invertebrates can be around 10,000 times smaller (or more) than the largest. Hence the difference between numbers and biomass. Studies of soil organisms have long noted that when plotting biomass on a logarithmic scale, there is much more equivalence between invertebrates of different size classes and this appears to be a consistent pattern (Polishchuk and Blanchard, 2019).

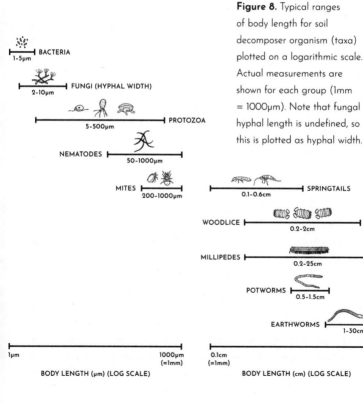

Figure 8. Typical ranges of body length for soil decomposer organism (taxa) plotted on a logarithmic scale. Actual measurements are shown for each group (1mm = 1000µm). Note that fungal hyphal length is undefined, so this is plotted as hyphal width.

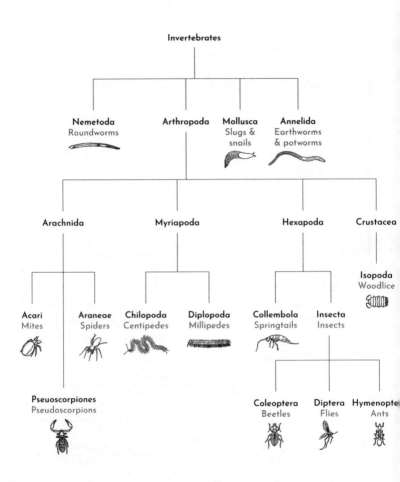

Figure 9. 'Tree of Life' showing the taxonomic relationships between the more significant members of the decomposer community in compost. This scheme is simplified for clarity and doesn't necessarily reflect current taxonomic terminology and organization. Several less significant types of invertebrates are not included.

Because scientists like to categorize things, they have done this with soil/compost invertebrates. Referring to body length, at the large end of the scale are the macrofauna (> 2 mm/ 0.08 in), such as earthworms, woodlice and millipedes. The mesofauna – for example, springtails, mites and potworms - are medium-sized (0.1-2 mm/0.004-0.08 in), while the microfauna (< 0.1 mm/0.004 in) are organisms like roundworms and single-celled protozoa.

Apart from roundworms, mites, springtails and potworms are also always numerous. An illustration of how all these different invertebrates are classified according to their evolutionary relationships is shown in Figure 9, a very much simplified 'Tree of Life'.

The compost food web

The next question is: what are all these invertebrates doing in the compost heap? The simple answer is that some of them are feasting on a diet of leaves and other organic matter (acting as detritivores). But in nature, answers are seldom simple. The compost heap harbours a very complex food web (Figure 10 on page 51). Based on detailed studies of soil invertebrate food webs, it is likely to have hundreds if not thousands of feeding links between invertebrate species (Digel et al., 2014). It is often referred to as a 'brown' food web, referencing dead plant material. This differentiates it from 'green' food webs, where animals (herbivores) are consuming living plants. In compost, various invertebrates feed in a variety of ways: many on fellow invertebrates and others on microorganisms, while yet others feed directly on organic matter. So, there is a pyramid of feeding relationships, although many are not fully known. At the base of the feed web is the input of organic matter. This is largely decomposed by microorganisms using their battery of enzymes. Some invertebrates will also chew the organic material directly, while others graze on the microbial flora growing on the dead plant material. After that, there is a cascade of invertebrates often feeding on each other – generally the bigger ones feeding on the smaller ones. In many ways this is no different from the 'green' food webs in larger-scale ecosystems. This is just on a smaller scale and with organic detritus as the source of energy and materials rather than actively photosynthesizing plants. There are some other types of interactions in the food web which we will come to later.

Plant litter feeders

There are relatively few invertebrate types that feed directly on the organic matter because most do not have the necessary enzymes. Even among those invertebrates that do feed directly on leaf litter, there will be differences in the extent to which they digest (decompose) the material as it passes through the gut. So, invertebrate faeces are also a substrate for further decomposition processes. The next sections look specifically at the contribution of specific litter-feeding invertebrates to decomposition.

Earthworms

These are arguably the champion invertebrate converters of dead organic matter in soil and compost across much of the globe. There is some tendency for greater earthworm species richness and abundance in temperate regions, but otherwise their distribution is dictated by climate and habitat characteristics (Phillips et al., 2019). It is unclear to what extent specific compost species of earthworms may also show differences between global regions.

Earthworms are frequently referred to as ecosystem engineers because of their impact on soils through their burrowing, mixing and transporting of soil material and consuming organic matter. They can also be thought of as keystone species because they have an impact on the ecology of soil

Figure 10 (right) Simplified food web of a compost heap. The breakdown of organic matter in compost involves a variety of organisms forming the decomposer community. This results in a cascade effect where energy and nutrients transfer between different 'levels' of consumers (first, second, etc.). Much of the nutrient content is retained in the compost heap while energy is ultimately lost as heat (see Figure 1, page 11). Transfer of energy between levels is by 'consumption'. This may be by microbial breakdown of organic matter or by direct feeding (for example, woodlice feeding directly on leaf litter; roundworms feeding on bacteria; ground beetles feeding on potworms).

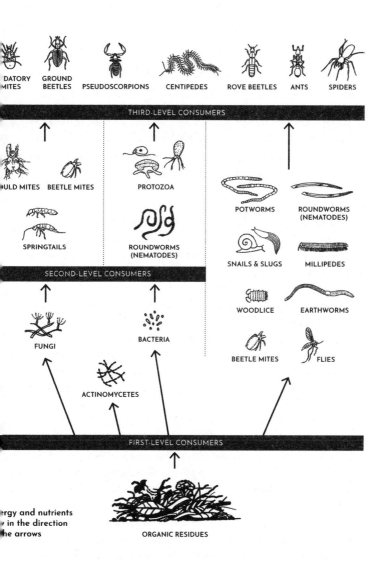

DATORY MITES GROUND BEETLES PSEUDOSCORPIONS CENTIPEDES ROVE BEETLES ANTS SPIDERS

THIRD-LEVEL CONSUMERS

ULD MITES BEETLE MITES PROTOZOA POTWORMS ROUNDWORMS (NEMATODES)

SPRINGTAILS ROUNDWORMS (NEMATODES) SNAILS & SLUGS MILLIPEDES

SECOND-LEVEL CONSUMERS

WOODLICE EARTHWORMS

FUNGI BACTERIA BEETLE MITES FLIES

ACTINOMYCETES

FIRST-LEVEL CONSUMERS

rgy and nutrients
in the direction
he arrows

ORGANIC RESIDUES

or compost disproportionate to their numbers (Brussaard et al., 2007). Charles Darwin was fascinated by their activity, not just in relation to breakdown of leaf matter but also in the way they work the soil. He recorded their habit of pulling leaf material into their burrows, where it is then consumed. He even undertook early experiments on the taste preferences of earthworms - noting that a variety of leaf types were consumed, but with a particular preference for wild cherry compared to lime and hazel (presumed to be *Prunus avium, Tilia* sp. and *Corylus avellana* respectively) (Darwin, 1888).

There has long been a debate about whether earthworms can secrete enzymes to digest cellulose in the leaf material they consume. Some studies have argued that the breakdown of lignins, cellulose and hemicellulose in these worms is largely down to bacterial and microbial symbionts in the earthworm gut. There is some evidence that earthworms do secrete a cellulase, but it seems it needs to combine with microbial cellulases for effective cellulose digestion (Fujii et al., 2012). It is also worth remembering that the term 'earthworm' covers a range rather than a single species - each with their own feeding niche and ecological preferences. Some are deep burrowing (endogenic) while others live in the surface layers of soil, especially in organic matter (epigenic). About half of the thirty UK species are found in gardens, but only two epigenic species have a particular affinity for compost heaps. The best known is the brandling or tiger worm - by virtue of its reddish colouration and striped appearance. This species, *Eisenia fetida*, along with a second, *Dendrobaena veneta*, have a liking for fresh organic material and a warm, moist environment. So, the physical state and structure of the compost material and the environmental conditions will affect the abundance of these two species of earthworm. As suggested above, they consume fresh organic matter and digest it through a combination of their own enzymes and those produced by their gut flora. Their contribution to the composting process is recognized through the concept of vermicomposting - which refers to artificially adding these species to hasten the commercial composting of a range of organic substrates. The worms themselves are a product that can, for example, be used as fishing bait. For a culinary adventure, earthworm meal has also been suggested as a human food. With a protein content of well over 50 per cent (dry weight) it is perhaps not surprising that there are several examples of humans consuming earthworms, from China to Africa to South America and even Western Europe (Sun et al., 1997).

Woodlice

Woodlice are very common in compost heaps – they wander through the air spaces created between leaves, plant stems, twigs and other material. They are also well known to most people in temperate regions, if not necessarily loved. This is evidenced by a wide range of vernacular names given to them matched by few other invertebrates. Many of the names have to do with cheese (for example, cheesy-bugs in the UK) and pigs (woodpigs in the UK and sow bugs in the US and Canada). Woodlice do consume dead plant material – and if you carefully peel away some loose tree bark, where they are often found, you may well spot a lot of granular material – woodlice faeces. Of course, other invertebrates also produce faeces, but those from woodlice are especially visible. These faeces add significantly to the breakdown products of decomposition, but that isn't the end of the story. The faeces themselves are then colonized by microorganisms, which take the decomposition a stage further. Because woodlice work the compost heap in a different way to earthworms, they complement each other rather than compete. Experimental evidence shows that a mix of woodlice and earthworms progresses decomposition more rapidly than when only one or other is present (Zimmer et al., 2005). As is often the case in the natural world, there is partitioning of the resources that different species exploit. Avoidance of competition can be a good strategy for species.

Millipedes and centipedes

If you find a long-bodied, segmented invertebrate species in the compost heap with many legs, there are two possibilities: centipede or millipede. How do you tell the difference? The 'obvious' answer, helped by a good lens, is that millipedes have two pairs of legs attached to each body segment, while centipedes only have one pair. Both are found in compost heaps but have very different lifestyles. At the front end, centipedes have a set of pincer-like fangs that inject poison. So, they are venomous and predators rather than detritivores. Small centipedes in compost heaps attack other small invertebrates, but – as I know from personal experience – the giant centipedes of the tropics have a very painful bite.

By contrast, millipedes contribute to decomposition. Millipedes found in compost heaps are broadly of two kinds. Their elongated bodies are either cylindrical in cross section or rather flattened. The cylindrical species can be referred to as snake millipedes and the others as flat-backed millipedes. The maximum size of a UK millipede is about 4 cm/ 1.6 in – unlike tropical species, which can be as thick as a finger and grow to about 30 cm/12 in. in length. The tropical species have been described as a good choice for a pet – if you avoid their noxious defensive secretions! In the compost environment, millipedes will contribute to shredding and decomposing leaf material, but generally they have a greater impact on decomposition of organic matter in the tropics where they often replace the activity of earthworms. In temperate regions, millipedes aren't solely restricted to consuming dead plant material and have been reported as grazing on fungal tissue (Crowther et al., 2011).

Mites

Soil mites are another group of soil animals that deserve mention in the context of litter breakdown. If you look at any quantitative data on invertebrates in soil or compost, mites may be small in size but they figure very prominently in terms of abundance (see Figure 7, page 43). The taxonomic name for them is Acari. These relatives of spiders (adult mites are eight-legged) vary widely in form and ecology. There are four main groups of non-parasitic mites. The one group especially connected with decomposition is known as Oribatid or beetle mites (Cryptostigmata). Under a microscope these small mites look like somewhat lumpen, slow-moving creatures. With plant organic matter all around them, they don't need to be especially athletic. They chew their way through litter and associated fungi. But, once again, there is the question of how they digest this litter material. Because of their small size, studies on their gut activity are not easy. So, while it is clear that they do use cellulases in digestion, it is again uncertain whether these are secreted by the mites or whether symbiotic microorganisms play this role. Both have been suggested.

Fly larvae

The insect order Diptera is a very large and diverse group with about 150,000 known species. Their larvae have a wide variety of feeding modes, from disposing of animal corpses, to parasitizing living animals and plants, to contributing to the decomposition of dead plant matter. One fly whose larvae live in organic detritus in compost heaps may be somewhat familiar in the UK. This is the St Mark's fly (*Bibio marci*), known as such because it emerges from its pupal stage in the soil or compost around St Mark's Day, 25 April. These rather large, shiny, jet-black flies are a common sight hanging in the air rather languidly over vegetation, often in large numbers. They are but one member of the fly family Bibionidae, many of which, when in the larval stage, scavenge leaf litter and other organic debris. Their contribution to this process is unclear, although it has been suggested that they are major decomposers of leaf litter in European forests (Frouz, 2018). Another fly family, the Sciaridae, are known as the dark-winged fungus gnats. They may be familiar to anyone who keeps houseplants. They love developing in houseplant pots with rather wet compost, often leading to numbers of adult gnats buzzing around the pot. What attracts them to houseplant compost also entices them into a damp compost heap of leaf litter. They contribute to leaf litter breakdown, although some are known to prefer more woody material. Because of their rather small size, they are unlikely to make a major direct contribution to decomposition in most situations but may make a tasty morsel for some of the predatory invertebrates that hunt through the compost.

Despite the apparent role of earthworms, woodlice, mites and fly larvae in leaf decomposition, some have argued that invertebrates have a limited role in directly breaking down leaf material. Such a view undoubtably underplays their role. Quite apart from direct consumption of leaf litter, invertebrates can also be thought of as facilitating microbial decomposition. Their chewing and cutting (comminution) of leaf material opens up the leaf tissues for speedier microbial decay. Comminution has two effects. In the first place, the outer surface of leaves is typically covered with a thin layer of wax – which does create a limited barrier to fungal and bacterial growth. This can be bypassed once the leaf is chewed. More importantly, comminution increases the surface area of the leaf substantially, making microbial attack possible from all sides. There have been numerous reported experiments where bags made of mesh of various sizes have been used to contain leaf litter. When placed in the litter layer of soil (or in a compost heap), the different mesh bags exclude different invertebrates, according to their size. Such experiments point to the fact that litter decomposition proceeds faster in the presence of invertebrates. A full mix of invertebrate species of all sizes generally has the greatest effect on speeding up decomposition, compared with microorganisms acting alone. Although the results of different experiments are quite mixed, it has been suggested (Seastedt, 1984) that invertebrates are responsible, both directly and indirectly, for about a quarter of decomposition, measured as loss of litter mass. Another study (Slade and Riutta, 2012) put the figure at 17–31 per cent. Yet some other studies in woodland have calculated (not measured directly) that invertebrates like millipedes, earthworms and bibionid fly larvae may consume the bulk of annual litter fall (Frouz, 2018). The point is, whichever figure applies, it is a substantial contribution to the process of leaf litter consumption. But a final point worth bearing in mind is that consumption does not equate with decomposition. As with woodlice, much of the consumed material then emerges as faeces, which are then decomposed further by microorganisms. However, this can also be thought of as a process of facilitating microbial decay.

Of course, invertebrates and microorganisms that directly feed on or decompose organic detritus are not the only members of the decomposer community in the compost heap. As you can see in the diagrammatic version of a food web in Figure 10 (see page 51), there are three other broad categories of organism present in compost. These include invertebrates that feed on microbes, invertebrates that feed on other invertebrates and microbes that 'feed' on invertebrates. Starting our travel from the detritus end of the food web, we first encounter microorganisms that feed on the dead organic material. These will then be grazed by a suite of invertebrates (microbivores) that harvest the bacteria and/ or fungi. So, we have an invertebrate feeding activity that seems to run counter to the requirements of decomposition – they are consuming the very fungi and bacteria on which decomposition so depends. It is possible to imagine invertebrate microbivores grazing like cows on a field of *Penicillium* hyphae and spores and on films of bacteria, which in turn could slow the colonization and abundance of both fungi and bacteria.

Microbial feeders

Which are the microbivores in a compost heap? The first point to make is that any invertebrates consuming dead plant organic material will incidentally consume fungi and bacteria as well. Earlier we discussed microorganisms on leaf surfaces (phylloplane), those found within plant tissues (endophytes) and others associated with plant roots (root microbiome). Microbes are part of normal consumption by invertebrates feeding on leaf and other dead plant material – and they contribute to the nutritional value of ingested plant material. Some macro-invertebrates (earthworms, millipedes, woodlice) that are not specifically labelled as microbivores will readily consume fungal mycelium in laboratory experiments (Crowther et al., 2011). But there are three groups of invertebrates that particularly select microorganisms for grazing.

Mould mites

There are some mite species that delight in a fungal diet. To make the point, we can divert for a moment from compost to an example of this relating to cheese making. When it is ripened in a traditional way, cheese rind is soon overgrown by a variety of fungi. This can include *Penicillium* and *Mucor*, which have been previously mentioned. A close look at the surface will reveal tiny movements – of somewhat hairy mites (that is, mites sprouting long setae) that wander the surface of the cheese eating the fungi. These are members of a subgroup of mites called the mould mites or Astigmata, which can even contribute to the flavour of the cheese. They are different from the Oribatid mites mentioned earlier – which belong to another subgroup, the Cryptostigmata. Although not the commonest of soil/compost species, astigmatid mites will be seasonally present on leaf litter feeding *Penicillium* and other fungi.

Although we have generally referred to a wide range of fungal species collectively, there is research evidence showing that invertebrates may be selective in the fungal species they consume. For example, they may avoid chitin-digesting fungi (chitinolytic) because of the possible danger of being themselves digested (Maraun et al., 2003).

Springtails

These small, six-legged invertebrates used to be classed as insects. Despite their six legs, which labels them as hexapods alongside insects, they are put into a separate group (Collembola) to reflect their evolutionary divergence. Their characteristic feature, and hence the name of the group, is a springing organ (furcula) under the abdomen. If you poke about in the surface layers of compost or soil, you may induce one of these springtails to perform a jump, which in relation to their length (< 1-6 mm/0.05-0.25 in) can be spectacularly athletic. However, beware of the 'one body form fits all' assumption about springtails. Species that live deeper in the soil lack this springing feature; there is nowhere to jump! What does link all springtail species is that they are primarily feeders on fungi - hyphae and spores - although they can also be opportunists. Feeding on bacteria, plant material and nematodes has also been reported (Sechi et al., 2014). There is evidence to show that springtails may have preferences for particular fungi, selected on the basis of odour (Hedlund et al., 1995). Although springtails are small, they (like mites) can be very numerous in soil and compost and their grazing activity could have an impact on microbial populations. Whether they have a significant impact on microbial populations may well depend on the balance of microbial and springtail abundance.

Despite their small size, under a lens springtails can be spotted quite easily if some compost is spread out over a pale background. A white tray with upright sides is most useful for springtail (and mite) spotting.

Roundworms

Another group of microbivores, the roundworms (or nematodes) are much harder to see. They are small but extremely numerous - see Figure 7, page 43. They have a worldwide distribution in soils but are especially abundant in northern temperate and sub-Arctic regions (van den Hoogen et al., 2019). This might also be the case in compost, but data are scarce. Although there are some large parasitic species, those that are free living are small, less than 2.5 mm/0.1 in in length, mostly around 1 mm/0.05 in (Ruppert et al., 2004). They are also rather featureless, accentuated by being either very pale or near transparent in colour. They have no legs, so can't walk through leaf litter. Rather they wriggle and glide on surface films of moisture. Roundworms cannot be seen without a microscope. It is possible to buy fairly cheap but effective digital models with built-in illumination that can be linked to a computer or mobile device. You will need to wash some leaf compost in a small amount of water and then, using a spoon – or, better still, a pipette – draw a drop of water from the bottom of the dish and place it on a glass slide (as used in microscopy) or plastic or glass equivalent. Use your microscope with either strong illumination from above, with a dark background, or a clear background with transmitted light. You should be able to spot the roundworms swimming in the drop of water. Rather than grazing, these nematodes will consume bacterial cells present in the water films through which they swim (Heal et al., 1997). Whether these microbivores have an impact on microbial populations is again unclear. One study suggested that significant effects were likely only when conditions are unfavourable for microbial growth (Schlatte et al., 1998).

Invertebrate predators and pathogens

Next, we will look at the invertebrates that catch or trap and consume other invertebrates. These are the hunters. We will also give brief mention of microorganisms that turn the tables on invertebrates as predators or disease-causing pathogens. They become the invertebrate killers.

At the smaller end of the size spectrum are predatory mites. Many of these belong to a third group of mites which we have yet to discuss, the Mesostigmata. These are not like the slow-moving beetle mites, or the Astigmata, which are encumbered by long 'hairs' (setae). This group of compost-living mites are long-legged and rangy – they move and pounce quickly, maybe on a nematode, a springtail or fly larva. Larger predatory invertebrates found in compost include centipedes, beetle larvae and adult beetles, all with fang or scissor-like 'jaws' with which to capture and consume their invertebrate prey. As discussed, in their armoury centipedes also have the ability to inject poison.

There are two important groups of beetles among the compost predators. The first are ground beetles. These are most familiar as the fast-running black beetles that sometimes scuttle away when you move a pot in the garden (*Pterostichus* spp.) The second group are rove beetles (Staphylinidae), which don't look much like typical beetles. They are rather long-bodied with a visible segmented abdomen – unlike most other beetles (for example, ground beetles), in which the abdomen is covered by hard wing-cases (elytra). One rove beetle in particular is familiar to many people and frequently encountered in summer and autumn, potentially in and around compost heaps. This is the rather fearsome-looking devil's coach-horse (*Ocypus olens*), which can bite and is a fast-moving and effective predator. The size of this beetle (their length is about 2.5 cm/1 in) is exceptional in the UK; most rove beetles are rather smaller but with a similar body shape. These larger hunters, along with their predatory larvae, are varied in their tastes: soft-bodied slugs, beetle larvae and fly larvae are all fair game.

A recent study attempted to take a holistic view of what determines overall taxon richness (that is, how many 'types' (taxa) of invertebrates are present) of leaf litter/soil invertebrates. The study was undertaken in single-species experimental plots (woodland monocultures) of 14 temperate tree species (Mueller et al., 2016). Although it did not deal with compost, it provided pointers to what affects species richness of invertebrates in decomposing leaf litter of different tree species. The authors considered 10 taxonomic groups of invertebrates and around 125 different ecological factors and generated a rather complex data set. When analysed, the data showed that each invertebrate taxon responded individually and had a particular optimum combination of factors. Nevertheless, there were several environmental conditions and resources that collectively were good predictors of total taxon richness when integrated across 9 invertebrate groups (excluding earthworms). These included light, presence of earthworms, phosphorus, nitrogen, calcium, soil pH and densities of microorganisms and the amount of organic matter. Each of these could have positive or negative effects on particular taxa. Overall taxon richness was particularly enhanced by availability of phosphorus, calcium and nitrogen. Most of these factors are covered in the various parts of this book.

Let's close this chapter with a quick look at the range of other organisms that attack invertebrates. For example, there are around 200 fungal species that are known to be predatory on roundworms (nematodes) (Hsueh et al., 2013). Most of these are species that grow saprophytically on plant organic material but that can also, in an opportunistic way, exploit nematodes as a rich source of nitrogen. Possibly the most fascinating of these are the nematode-trapping fungi. These fungi have developed remarkable hyphal structures to ensnare roundworm (nematode) prey. They may be sticky nets of hyphae, or noose-like hyphal rings that, when triggered, suddenly swell with an influx of water into the cell and 'strangle' the hapless nematode. These fungi have developed a sensitivity to particular signalling chemicals used by nematodes (ascarosides) so that the hyphal structures are grown only when the fungus is short of nitrogen and in the vicinity of an active nematode population. Scientific studies suggest that these types of fungi can have a tangible impact on nematode populations in soil or compost (Hsueh et al., 2013). Other fungi and bacteria affect invertebrates in other ways – they cause disease. Many different fungi have been shown to infect insects, and their asexual spores are known to survive well in soil. In the warm, damp environment of a compost heap, it is likely that members of the compost invertebrate fauna will periodically become infected and consequently die. However, there appears to be little evidence of any kind of mass mortality from this cause in compost. Other potential invertebrate killers in compost heaps include several naturally occurring species of roundworms (nematodes) that infect and kill a variety of insects and slug species. These have now been developed into commercial biological control products for pests such as fungus gnats (sciarids) found in mushroom compost and compost heaps.

9 / PHYSICOCHEMICAL ENVIRONMENT AND DECOMPOSITION

There are a range of conditions that either speed up or slow down the rate of rotting in a compost heap, some of which have already been hinted at in previous chapters. What follows is a more comprehensive overview – in the context of a northern temperate climate, with the focus on northern Europe, but that is relevant to other climatic zones as well. It is important to stress that all the various factors listed here contribute to making decomposition work – or, in ecology parlance, any of them might become a limiting factor on the rate of decomposition. They all need to be 'right' for rapid breakdown of organic matter. Once again, this is only a partial review of a complex set of conditions that act separately but will also interact. The emphasis is on major environmental factors; seemingly more minor ones are given passing mention elsewhere in the book.

Starting with the nature of the compost itself, what is it about the composition of compost that makes it decompose faster or slower? As already mentioned, all living organisms require a balance of nutrient elements to build cells and tissues. Generally, the more nutrient rich the compost, the faster it is likely to decompose. Going back to the example of tree leaf litter: we already know that, prior to leaf fall, tree leaves lose a lot (on average, about 50 per cent) of their nutrients, which are withdrawn (resorbed) back into the tree. If we were to select one element that is likely to be in especially short supply in relation to decomposition, it would be nitrogen. Here there is another paradox. In the atmosphere, nitrogen is all around (about 78 per cent of air by volume) and yet in ecological systems nitrogen can often be limiting to growth and reproduction. Having said that, in natural ecosystems (and with some types of crops), there is a way in which nitrogen can be drawn down from the atmosphere by certain types of nitrogen-fixing bacteria, which capture nitrogen gas and incorporate it into biologically useful molecules. Some of these bacteria are symbiotic (*Rhizobium* genus) and typically found in root nodules of plants of the pea and bean family (Fabaceae). They exchange nitrogen for sugar from the host plant. In addition, there are free-living soil bacteria that can perform the same biological trick (for example, *Azotobacter*). Despite such biological processes of nitrogen fixation, gardeners add additional nitrogen fertilizer to the soil in flower or vegetable beds to make their plants grow faster. Some of this nitrogen then finds its way to the compost heap through dead vegetation and some may eventually be returned to the garden when mature compost is added to soil.

Nitrogen

So, some more thoughts about nitrogen in compost heaps. First, remember that nitrogen is a key component of proteins – no proteins, no new cells. For decomposers there is the problem that the nitrogen content of plant tissue is generally much lower than that of animal or microbial cells. What about the nitrogen content of leaf litter? Most of the nitrogen in dead leaf cells will be present in what remains of large organic protein molecules. Many of these will be easy for microbes to decompose and some will be consumed and digested by litter-feeding invertebrates. Through a range of digestions and biochemical conversions, this nitrogen then cascades through the food web as it is used by a variety of microorganisms and invertebrates. Typically, it is incorporated in new protein molecules – but some may be released as inorganic nitrate (NO_3^-), which is easily soluble in water and so can migrate through the soil or compost in moisture. Inorganic nitrate will be quickly absorbed by microorganisms (and then indirectly by invertebrates grazing on microorganisms) and reincorporated into protein. In this form, the nitrogen is trapped in cells (referred to as immobilized or unavailable) and so cannot be used by plants if the compost is added to soil. Nevertheless, the immobilized nitrogen in microorganisms and invertebrates is eventually released when these organisms die and are decomposed in the soil and can then be taken up by plants. Even at this point, plants will still be in competition with microorganisms for this 'available' form of nitrogen. Useable forms of nitrogen are a precious resource in both soil and compost.

As noted previously, dead plant material will generally be poor for nitrogen – so there will always be a big race to grab what is available. However, some insoluble proteins in animal tissue contain valuable nitrogen (for example, keratin in skin and collagen in connective tissue), as do the cell walls of fungi, which have a structural, nitrogen-containing polysaccharide/polymer called chitin. All these materials must be broken down (depolymerized) with specialist enzymes (proteases for protein and chitinases for chitin). Some organisms arriving late to the process of decomposition can still thrive if they have the right enzymes to break down these difficult molecules.

If the waste plant organic matter is woody, nitrogen is even more of a problem. Woody tissue is dominated by cellulose and especially lignin. In general, there is an inverse relationship in organic matter between nitrogen content and lignin and other recalcitrant polymers (Heal et al., 1997). In plain English, this means the more lignin, the less nitrogen.

Fresh wood has a thin layer of living tissue on the inside of the bark, and an active outer layer of wood where the living cells contain nitrogen and make a reasonable substrate for decomposition. But in heartwood or twigs and branches that have been dead for some time, nitrogen limitation is a real issue. One way to illustrate this problem comes from woodboring beetles (saproxylic species), of which the well-known furniture beetle and deathwatch beetle are examples. These and other beetle larvae choose dead wood in which to develop. The lack of nitrogen makes this a slow process, to the extent that such larvae can take several years to develop to adulthood. That is how long it takes for them to chew through enough wood to extract enough nutrients (especially nitrogen) to mature. When they are ready to emerge from a pupa, they burst through the wood surface as adult beetles in search of new wood to colonize. When you see holes made in wood by furniture beetles, the adult beetles have already departed and flown. You may also notice little piles of 'sawdust' (frass), which are the partially digested wood they have chewed through to extract the very low amounts of nitrogen. The beetle larvae need to chew their way through a lot of wood! In more natural environments, decaying wood will be subject to attack by a range of woodboring (saproxylic) invertebrates, especially woodboring longhorn beetles (Cerambycidae family). Although such beetle larvae are unlikely to be active in compost heaps (as opposed to a wood pile), this does highlight the nature of the predicament for organisms that decompose wood.

Although insect larvae have some impact on wood decay, the main agents of decomposition are fungi, which face the same problem of nitrogen shortage. But they have a clever strategy to overcome this problem and so can work the wood faster. Their trick is to have an ever-expanding mycelium front of hyphae growing through the wood. This mycelium then dies back in the wood that has already been decayed. But before this happens, the nitrogen is transferred (translocated) to the advancing hyphae spreading through the fresh wood. This is a good system of recycling that ensures the fungus is not short of nitrogen while scavenging what limited extra nitrogen is available as decay moves forward. A similar strategy is adopted by gradually expanding rings of mushrooms in a field – with just the edge of the circle producing the fruiting bodies.

Nitrogen levels form the basis of a simple measure of how decomposable the organic material in the compost heap is likely to be. This is known as the carbon:nitrogen ratio (C:N). As an example, if there is 100 times as much carbon as nitrogen in compost material, the ratio will be 100:1. The actual value will vary widely, depending on the type of material in the compost heap. As a rule, the greener the material, the lower the C:N ratio (that is, a ratio of 50:1 is lower than 100:1). For example, autumn leaves have a C:N ratio of 30-80:1, while grass clippings are 15-25:1 (Cornell Waste Management Institute, 1996a). The optimum for compost heaps is often quoted as about 30:1. Less than that is undesirable (and generally unlikely in garden compost with no added food waste) because of possible release and waste of surplus nitrogen as noxious ammonia gas.

So how does nitrogen get recycled in compost? The starting point is when nitrogen-containing material is added to the compost heap. As noted before, this will be mainly in the form of protein but also other molecules like DNA. In this organic form, the nitrogen cannot be used directly by organisms to grow – it generally needs to be converted to smaller, soluble inorganic molecules (or ions) (see Figure 11, page 67), although it is now known that some smaller organic molecules containing nitrogen can be taken up directly by microorganisms (Geisseler et al.,

2010). This process of conversion from organic to inorganic nitrogen is referred to as mineralization. A wide range of soil bacteria can perform the first stage of this process – from protein molecules to ammonia (NH_3) or ammonium ions (NH_4^+). This stage is known as ammonification and is an energy-yielding process for bacteria – which is why they do it. The next stage, known as nitrification, is down to two specialized genera of bacteria (*Nitrosomonas* and *Nitrobacter*), which respectively convert (oxidize) the ammonia into nitrite (NO_2^-) and then to nitrate ions (NO_3^-) in a two-stage process that is also energy-yielding. This was long thought to be the only way for converting ammonia to nitrate. More recently, *Nitrospira* bacteria were shown to be capable of completing both steps of the oxidation process (Cáceres et al., 2018). Why is this nitrification stage really important? Although microorganisms can readily utilize ammonia as a source of nitrogen for synthesizing protein (Geisseler et al., 2010), terrestrial plants typically do not use ammonia directly in their metabolism. For them the main source of soluble nitrogen in the soil is in the form of nitrate ions, which can readily be absorbed and metabolized. Just one more point related to nitrification. As already noted, ammonia gas may be released from compost that is particularly rich in organic nitrogen. The explanation for this is that nitrification is very much the rate-limiting step in nitrogen mineralization – so if the wrong conditions in the compost slow nitrification, ammonia gas will escape and nitrogen will be lost from the soil/compost (Figure 11).

There is another way in which nitrogen released during decomposition can be lost. Under anaerobic (low/no oxygen) conditions, some of the newly formed nitrate ions may be returned to the pool of atmospheric nitrogen by the process of denitrification (Figure 11). This is unlikely to be significant in compost heaps, though, unless the organic waste is compacted and wet with little exchange of air.

Of course, in practice it is not possible to make a single assessment of the C:N ratio in compost – it will vary throughout the compost. But there is a take-home message from this discussion about nitrogen. Compost heaps are more likely to contain too little nitrogen (high C:N ratio) rather than too much (low C:N ratio). So, if you add some nitrogen to your compost

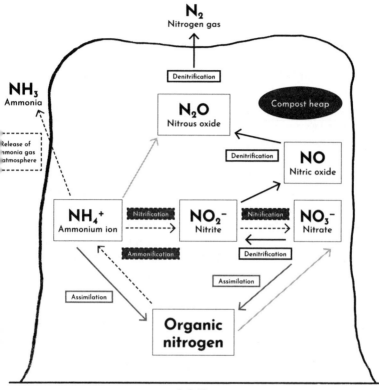

Figure 11. Nitrogen decomposition cycle. Dashed arrows: process of nitrogen mineralization – from organic to inorganic nitrogen. Solid arrows: mineralized nitrogen (nitrite and nitrate) can be lost by denitrification to a gaseous form (nitrogen). This occurs under oxygen-free (anaerobic) conditions only and so is unlikely to be a major process in compost. Pale arrows: both ammonium ions and nitrate ions can be absorbed by plant roots and used in growth (assimilation). This can happen following the addition of compost to garden soil. Very pale arrows: two processes that are unlikely to be significant in compost.

heap, plant organic matter should decay faster. This may be done through diversifying inputs into the compost heap - for example, green waste or kitchen waste, which are both likely to be more nitrogen rich. Manure and urine (urea) are especially good accelerators for composting - but must be added sparingly. Urea is an accessible form of nitrogen for microbial decomposers and they will respond quickly. Compost accelerators (or activators) that can be bought commercially claim a variety of ingredients, but especially a source of nitrogen. This might include commercial fertilizers like ammonium nitrate/sulphate or urea.

Other elements

It is not just nitrogen that can be in short supply in the compost heap. If you think of fertilizers used in bulk quantities in farming (macronutrients), three elements figure: nitrogen (N), phosphorus (P) and potassium (K). Each of these can be quantified in terms of C:P and C:K ratios - but again, this is beyond the scope of garden composting. However, the addition of small amounts of a general NPK fertilizer (or organic alternatives) may have additional benefits over and above that of nitrogen alone. As we know, compost varies in composition, so this can never be an exact science with compost heaps or bins in a garden setting. Generally, the most benefit from added nutrients will relate to nitrogen - but without adding too much.

Organic material

As well as providing crucial elements for cell construction, compost material also needs to supply a source of energy to the decomposer community. One thing that will be plentiful in compost heaps is polymer carbohydrates. A number of these are hard to decompose (for example, cellulose and lignin), while others, such as sugars and the plant storage product starch, are easier to digest or decompose. Because materials like cellulose are present in such copious quantities, organisms that have enzymes to break down these large molecules into their sugar constituents will not be short of energy. A skeletonized leaf demonstrates the initial breakdown of the easy bits of leaf tissue, based mainly on cellulose (such as leaf mesophyll cells) and a longer time period required to decompose the lignin-containing vein tissues.

A compost heap contains a range of plant residues - the nature of which will vary from garden to garden and season to season. This makes it tricky in any given situation to assess how different vegetation types impact on decomposition rates. However, in most gardens, leaf litter is a major input - either from the garden's own trees or blowing in from adjacent gardens. In the autumn this may be the dominant input of organic material into the compost heap. So, what difference does leaf type make to decomposition rate? Some of the tree species in gardens are likely to be native, but non-native species from across the globe are equally likely to be present. Gardens can be biodiversity hot spots! Are the local populations of microbes and invertebrates equally able to deal with these more exotic offerings?

A useful UK study of decomposition rates of senescent foliage of 125 vascular plants was published several years ago (Cornelissen, 1996). Many of these species were native, but others were invasive or non-native ornamentals. To cut a long story short, some broad generalizations can be drawn from this investigation. Perhaps the greatest contrast was between leaves from deciduous trees and foliage from evergreens. The study included several conifer species and other evergreens such as *Rhododendron*, which were all generally slow to decompose. Overall, deciduous species decomposed twice as fast as evergreens. This indicates that, for example, needles from a conifer Christmas tree or prunings from a conifer hedge are likely to decompose rather slowly in a compost heap. Herbaceous (non-woody) species decomposed faster than leaves from trees and shrubs. Similarly, leaves from herbaceous species decomposed faster than senescent (dry) grasses - although this might not be true if the comparison was with fresh grass cuttings. There was also a broad study made of decomposition rates of foliage from nine different plant families. Leaves from the family Fagaceae (which includes oak and beech) were the slowest to break down, and results were similar for conifers (family Pinaceae - pines, firs, spruces and so on). Decomposition rates were faster in the family Fabaceae (peas and beans), for example, and more so the Asteraceae 'daisies'. In some ways it was surprising that the Fabaceae were not the fastest to decompose, because the nitrogen-fixing bacteria in their roots means they tend to have high levels of nitrogen in their tissues – making them a useful addition to the compost heap.

Another observation in the study was that brown leaves decomposed more slowly than yellow or green leaves. The thinking here is that green (and to a lesser extent yellow) leaves have retained more nutrients at leaf fall than those that are brown, and brown leaves are likely to have a higher content of difficult molecules such as lignin and tannins. The study included various trees and shrubs that had leaves still classed as green at leaf fall – for example, alder (*Alnus glutinosa*), ash (*Fraxinus excelsior*) and privet (*Ligustrum vulgare*). These species decomposed rather quickly.

Despite the study's inclusion of a range of native and non-native species, there were no specific conclusions relating to differences in decomposition between these species. Another study (Jones et al., 2019) included only four tree/shrub species; here, English oak (*Quercus petraea*) and ash (*Fraxinus excelsior*) were compared with sycamore (*Acer pseudoplatanus*), which is possibly native, and rhododendron (*Rhododendron ponticum*), a non-native invasive species. The non-native rhododendron decomposed slowly, but no slower than the oak, whose leaf litter is known to decay slowly. The ash and sycamore decomposed rather faster. Other individual studies have suggested that litter from exotic woody species decomposes faster than native species (Ashton et al., 2005). A worldwide review of the literature also concluded that non-native invasive species (woody and non-woody) decompose faster, typically because of lower C:N and lignin:N ratios (Liao et al., 2007). These studies suggest that a mix of native and exotic leaf litter may even promote faster rather than slower decomposition in compost, although decomposition of many evergreen species will be slower.

Already discussed are the problems for invertebrates of digesting difficult substrates like cellulose or lignin. But there is another side to the story. Are there substances in leaf litter or other organic debris that are toxic or in some other way discourage invertebrate feeding? In plants a distinction is made between what are called primary and secondary metabolites. The former relates to those metabolic (biochemical) processes which are essentially common to all plants. Secondary metabolites are those that are individual to particular plant species. These metabolites can be encountered as, for example, the flavours of culinary herbs, or the scent of foliage or flowers. It is argued that many of these secondary metabolites have a defensive role against herbivores and plant pathogens. This then raises the question of what effect they will have on the decomposer community when they form part of the spectrum of organic molecules that decay. For example, we have previously mentioned the complex polymers of tannins. These are secondary metabolites found especially in tree bark but widely occurring in many plants. It is known that these tannins can bind to soluble proteins making them insoluble. It was long thought that this would be an issue for detritivores feeding on tannin-rich plant litter - as would be true of oak leaves. The argument was that tannins would inactivate digestive proteins in the invertebrate gut and 'give them a bad tummy'. However, experimental evidence doesn't support such a drastic effect - it seems that decomposer invertebrates are well versed in dealing with the problem of tannins.

Another well-known example of a secondary metabolite that may pass through a compost heap is a substance called juglone, which is especially abundant in the leaves of trees belonging to the walnut family (Juglandaceae). Interestingly, this substance inhibits the growth of other plant species rather than being targeted against pests or pathogens. Technically it is described as an allelochemical. The growth of a number of garden plants - for example, tomato, potato, apple, cucumber and ericaceous species such as rhododendron - is affected by juglone in the

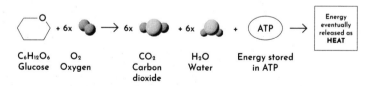

Figure 12. Release of energy from glucose by cellular respiration (respiratory metabolism). Figures indicate numbers of molecules involved. Energy stored in ATP is then used by the cell in the synthesis of new molecules.

soil. A recent review of the effect of plant secondary metabolites on nutrient cycling listed many possible effects on both invertebrates and microorganisms but quoted often contradictory outcomes of scientific studies (Chomel et al., 2016). So, the jury is still out.

The end result of decomposition isn't the complete disappearance of organic material, but the compost heap does shrink in volume. Much of the carbon will have been lost as carbon dioxide – the result of breakdown of lignocellulose to sugar molecules, which are used as a source of energy within cells. Because it is the breakage of the bond structure of the sugar molecules (the linkages between carbon, hydrogen and oxygen) that releases energy, the elements remain and recombine into 'waste' molecules of water (H_2O) and carbon dioxide (CO_2) (Figure 12). The water remains within the depths of the compost, while the carbon dioxide diffuses out and dissipates into the atmosphere. Because of their extreme mobility in the atmosphere, those carbon dioxide molecules may well be absorbed in some distant rain forest as a feedstock for photosynthesis, before entering the decomposition cycle again in a more exotic location.

Humus

After a year of decomposition in temperate regions, about one-third of the carbon in plant organic matter remains in soil (Stevenson, 1994). If you dig out material from the base of your compost heap after twelve months or more, what will you see? It is likely to be dark-coloured material – often nearly black. The organic component may well include the remains of some particularly difficult material that is most likely very lignin rich. However, it is a mistake to think that decomposition is only about the breakdown of materials. Synthesis also occurs. The term 'humus' is commonly used to refer to well-decomposed plant material. More accurately, humus will be a mixture of recalcitrant material yet to decompose as well as new humic substances synthesized by microbial activity during decomposition and possibly during passage through the gut of invertebrates (Frouz et al., 2011). As is true with lignins, there is no single molecular structure for humic substances – they are also large and complex molecules. One of the key fractions is labelled as humic acid, which doesn't mean that this is a single type of molecule – there are several acidic humic molecules.

In processing terms, humic substances are mostly stable end-products of decomposition. It may take many years for further and very slow change.

What about the effect of variations in physical factors on decomposition rates? There are four important ones: oxygen, water, temperature and acidity/alkalinity (pH). Although discussed separately, it is again important to stress that there will be a degree of interdependency between these factors. A change in one will affect the others.

Oxygen

This gas is a major component of 'normal' air (about 21 per cent oxygen by volume) and is an essential requirement for decomposition. Why is air needed in the compost heap? Simply because all the biota typically responsible for effective decomposition have cells and tissues that are continuously biochemically active. They are constantly transforming and synthesizing molecules as part of their growth. These processes require a source of energy to drive them – so carbohydrates (sugars) and other substrates are continuously broken down enzymically to release energy and carbon dioxide (cellular respiration – see Figure 12). It is somewhat akin to a factory that needs energy as well as raw materials to drive the machines to make a product. In the cell these two processes combine under the banner of metabolism. And metabolism requires oxygen. Another partial analogy would be to think of a fire burning coal or wood to release energy (in that case, heat). A constant flow of air is required to keep the fire going. No oxygen means very little composting activity. Breakdown of organic matter is largely an aerobic (oxygen-requiring) oxidation process.

Hence, the compost heap requires steady movement (diffusion) of oxygen right into the middle of the heap. If all the organic material is packed too tightly, then this doesn't happen efficiently, and decay slows down or stops. With a good mix of organic materials such as leaves, dead plants, twigs piled loosely in the heap or bin, all should be well. But just to make sure, some gardeners make a habit of turning the compost. This loosens things up and also ensures that all the organic material is mixed and adequately exposed to favourable conditions – to get even decomposition.

There are two further points to add. As noted above, there may be interaction between the four factors listed. For example, having a loose compost structure or turning the compost regularly are both good ideas in principle. However, they may contribute to a drop in temperature or drying out the compost more quickly so that it may become too dry or the temperature too low for effective decomposition.

The final word in this section slightly qualifies a statement made earlier about decomposers needing oxygen. That is not entirely true, because some bacteria can still operate in the absence of oxygen. Their metabolism works in anaerobic conditions - no oxygen. There will be situations in which oxygen is in short supply and anaerobes come into their own. But this is a much slower and much less effective process that results in only partial decomposition.

Water

Perhaps as an environmental factor water is better referred to as moisture. Like many things in life, you can have too much or too little of it in your compost heap. Too little means that many of the organisms on which decomposition depends cannot function. It's quite simple - they dry out (desiccate) and either die or go into some kind of inactive (dormant) state until moisture levels are restored. This is very understandable in the case of delicate structures like fungal hyphae. But invertebrates can also dry out. As already mentioned, nematodes are reliant on a film of moisture in which to swim and forage. Woodlice are also vulnerable to moisture loss: unlike insects, these crustaceans do not have a waterproof waxy layer in their exoskeleton - which is why they prefer damp locations such as under bark, under leaf litter or the interior of compost heaps or bins.

Essentially, all cellular metabolic activity occurs within a watery medium - which is why it is so crucial for life. So, some water is good. But more water is not necessarily better. You can have too much of a good thing - and here the point about interaction between factors becomes especially relevant. In a compost heap saturated with water, oxygen will not penetrate so easily. Oxygen can diffuse slowly through still air - for example, to replace carbon dioxide generated by decomposition. But oxygen diffuses much more slowly through still water - anything from x5,000 to x10,000 slower (Cornell Waste Management Institute, 1996b).

Problems arise when, for instance, a quantity of fresh grass from the lawn mower is added in one go to the compost. As this starts to decompose it forms a rather compacted and wet mush of grass. Remember that microorganisms release water as well as carbon dioxide as a byproduct of metabolism. Both these waste products will hinder further decomposition by aerobic microbes and invertebrates. This is why such a lump of grass can seem little changed (apart from in colour) after some months in compost. The trick is to ensure that the grass is well mixed with other organic matter. This will open it up to air circulation and all should be well.

At the other end of the moisture scale, too little moisture will seriously restrict decomposition. Moisture availability can be measured in a variety of ways. A study by Meentemeyer (1978) expressed it as actual evapotranspiration (the amount of water evaporating from surfaces and plants) and reported that it was a strong climatic predictor of decay rates of plant litter at five different locations in Europe and the United States. On this broad geographical scale, higher rates of evapotranspiration (primarily the combined effect of water availability and temperature) correlated with faster litter decomposition. This is the pattern to be expected in hotter climates with sufficient rainfall.

Temperature

This is the third environmental factor to consider. Everyone is aware that temperature, changing as it does between seasons in temperate climates, contributes majorly to driving seasonal patterns of plant growth and animal activity. The reason for this is down to basic chemistry - and by extension, to metabolic processes. As a very broad generalization, the rate of chemical reactions more or less doubles with every ten degrees rise in temperature (expressed by the abbreviation Q10 = 2). While there is wobble room around this statement, it is reasonable to argue that a warmer compost heap will also be one that decomposes faster - all else being equal. If you probe a compost heap with a thermometer, it will typically be warmer on the inside than on the outside. The difference may be only a few degrees but can be substantially more, depending on the time of year, the nature of the compost pile and the stage of decomposition.

Figure 13 illustrates the very dramatic differences in temperature possible in large-scale and in domestic composting.

At the top end of the temperature scale is what is known as 'hot composting'. High temperatures can be generated within the compost in the right conditions to promote fast decomposition. Not only that, but such high temperatures have other benefits, such as killing weed seeds and disease-causing organisms. Starting with the mesophilic phase, microbes adapted to relatively low temperatures work on the 'easy' molecules within the pile of fresh organic matter. They can become very active very quickly – but why does the temperature rise? In our previous discussion of metabolism, it was noted that the process of energy release results in waste products of water and carbon dioxide molecules. But there will also be the release of waste heat (Figure 12, see page 71). You can think about this in terms of your own activity. When you engage in

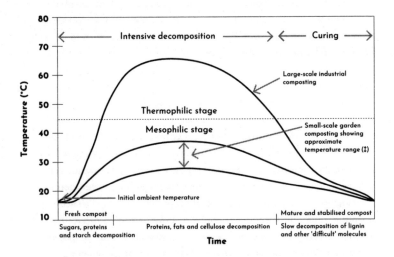

Figure 13. Temperature changes during commercial versus domestic/garden composting. Heat released through the metabolic activity of the decomposer community in the compost.

physical exercise your muscles do work and require an energy source to drive them – which may be the breakfast you have eaten that morning. However, once a muscle fibre has contracted, that energy is degraded to waste heat and you feel hot. As is drummed into everyone in school science, energy cannot be created or destroyed, only turned from one form to another. In this case, the energy you take in as sugar in breakfast cereal is ultimately released as heat, initially to keep your body at a stable temperature but ultimately to escape to the air around you. Similar things are happening in the compost. If the heap or bin is large and/or well insulated, heat escapes slowly and the internal temperature starts to build. Once it is over about 45°C/115°F, conditions reach the thermophilic phase. At such high temperatures only certain species of microbes that are adapted to high temperature (thermophilic) can survive and their metabolic activity contributes to further heating. So, at these high temperatures there is large reduction in microbial abundance and diversity: fungi are almost eliminated during the thermophilic stage (Ryckeboer et al., 2003). Meanwhile those invertebrates that are nimble enough will move to the cooler outer regions of the compost, while others succumb to the rising temperature.

When the majority of the key structural molecules (such as cellulose) have largely decomposed, the supply of both chemical energy and nutrients begins to fall. Instead of being in the plant litter, the nutrients are now largely locked up (immobilized) within the decomposer organisms. Of course, these organisms will eventually die and trigger further decomposition processes. In the cooling phase – known in composting terminology as the curing or maturation phase – the compost is back to the mesophilic temperature range and further decomposition occurs much more slowly. This is because of not only the lower temperature, but also the lack of easy energy-providing molecules and the type of difficult (recalcitrant) material that is left. With the lower temperatures, heat-intolerant microorganisms and invertebrates can repopulate the compost.

To have a chance of reaching the higher end of the compost temperature range, it is important to have a large volume of compost, high nutrient input (for example, lots of mixed kitchen waste, manure and so on), summer ambient temperatures and/or a very well-insulated compost bin. At the same time, there needs to be sufficient air circulation in the compost to fulfil the requirement for oxygen and the dispersal of carbon dioxide. There are some commercially produced insulated garden compost bins that are advertised to reach a temperature of 40-60°C/105-140°F and should therefore achieve much faster decomposition rates than you would expect in a traditional compost heap. However, most garden compost heaps will not go through anything like such extreme heating and decomposition. In a normal garden environment, the trajectory of decomposition will be much less extreme and more drawn out in time. In wintertime, the temperature of the compost may be close to ambient air temperatures, because little decomposition means little heat generation. A study by Imperial College in London (Smith and Jasim, 2009) of 64 household compost bins found that in the real world of garden compost, the majority of recordings of garden compost temperatures were in the range 20-30°C/68-86°F in summer and lower in winter. In general, many temperature readings were only just above ambient temperatures, although a small number of bins achieved temperatures beyond the top end of the mesophilic range (>45°C/115°F).

Acidity/alkalinity

So far we have made no reference to compost pH. Decomposition rates can be affected by acidity or alkalinity. This is measured on a scale of 1 to 14, with neutral at pH 7. This scale is logarithmic (rather that arithmetic) – so it is worth remembering that a change by one pH unit means a tenfold change in acidity/alkalinity. Individual decomposer species have a preferred pH in which to operate. The nature of the organic material added to the compost will have an influence on pH. However, unless there is thorough mixing of the organic matter, you may also end up with a patchwork of material each with different conditions on a microscale, including that of pH. But why should the pH be variable? Decomposition processes can release organic acids. In the early stages of decomposition, simple organic acids like acetic (ethanoic) are a product of microbial activity. Later, decomposition products include humic acid, as mentioned earlier. Published information indicates that humic acid is a weak acid with a pH of 6 – on the acid side of neutrality (Ali and Mindari, 2015).

Even carbon dioxide generated by microbial and invertebrate activity, when dissolved in water, can take the pH down to a weakly acidic pH 5.5. Countering acidity in compost is the calcium content of the leaf litter or other organic and soil material. To put it simply, the calcium helps to neutralize the acidity, although a chemist would describe what goes on in rather more complex terms. The calcium content of leaf litter will generally be affected by the nature of soil from where the plant grew: acid or alkaline.

Alkaline soil can also be described as 'base rich' – the terms are essentially used interchangeably. As the main bases in many soils are calcium compounds, primarily in the form of chalk, such soils can also be described as calcareous. Plants growing in alkaline/base-rich/calcareous soils will typically have higher calcium concentrations in their tissues than those growing in acidic (base-poor) soils (Reich et al., 2005). Some plants have an apparent preference for base-rich, chalky soils. These are referred to as calcicole plants. Others are confined to acid soils (calcifuge). In between these more extreme preferences are many species that have a broad range of tolerance to variations in soil pH. In the UK, for example, tree species such as sessile oak, Scots pine and birch will frequently occur on acid soils and so might have low levels of calcium in their foliage. Other species like ash, field maple, elder and field elm are frequently found on more alkaline soils and so may have higher levels of calcium in their leaf tissue. There are also species such as beech that can tolerate both acid and alkaline soils. The origin of leaf litter can therefore influence the pH of the developing compost. In a broader context, it is also known that litter composition can vary between plants of the same species, depending on specific site conditions (for example, Sanger et al., 1998).

Perhaps the main reason pH is important in relation to decomposition processes is its significant influence on the microbial community. Setting aside the impact on individual microbial species, there are some general effects that have been confirmed by various studies. One of the most convincing was based at the Park Grass experimental plots at Rothamsted Research in Hertfordshire, UK. The data come from a field experiment that has been running since 1856: grass plots with a range of different conditions (referred to as 'treatments'), including manipulation of soil pH, with pH values between 3.3 and 7.4 in different plots. Studies on the microbial populations in these plots revealed that the balance between bacteria and fungi changes with increasing soil acidification. At the low end of the pH range (acid), relative abundance shifted very much in favour of fungi. This makes clear that fungal decomposers are favoured by acid conditions in compost, while bacteria are much more active if the compost is neutral or edging towards alkaline (Rousk et al., 2011). Important decomposition processes driven by bacteria, such as the conversion of organic to inorganic nitrogen (mineralization), can be greatly affected by pH. As we noted earlier, nitrogen in protein is released first as ammonia and then converted in two stages to nitrite (NO_2^-) and nitrate (NO_3^-). The bacteria involved in this ammonia-to-nitrate conversion show a reduction in nitrification as the pH of soil/compost becomes more acid (Zhang et al., 2017). So, there is clear evidence that compost pH can have a significant impact on essential microbial activity.

Another effect of compost pH is on the accessibility of nutrients by plants growing in a compost-based soil. Although the chemistry is quite involved, in simple terms the pH can affect the solubility of chemical compounds of a range of key elements previously listed. This is illustrated in Figure 14, where the wider the bar for a particular element, the more accessible it is for uptake at a particular pH. So that, for example, in a soil or compost with a moderately alkaline pH, say pH 7.5-8, important nutritional elements such as phosphorus (P) and iron (Fe) are less available in soluble form. As such, those organisms unadapted to base-rich environments may experience nutritional problems. On the other hand, in more acidic soils a variety of metal elements can cause toxicity problems for plants. Note, for example, the high levels of aluminium (Ai), which can be toxic to unadapted species in compost with low pH – although this is unlikely to be an issue at 'normal' compost pH. Subject to how much of a particular element is stored in the soil or compost, the best availability for most elements is about pH 6.5-7. At that pH, nutrient limitation is unlikely in a soil well provisioned with accessible nutrients.

What about the impact of compost pH on decomposer invertebrates? The most quoted effect is on earthworms and their allies. Earthworms

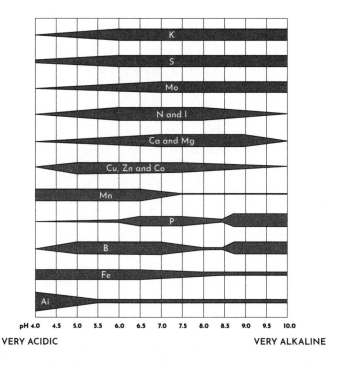

Figure 14. The effect of soil/compost pH on solubility of compounds of different nutrient elements and hence their availability for uptake by plant roots.

are sensitive to soil acidity and are much more frequent in alkaline (base-rich) soils (Reich et al., 2005). Depending on the earthworm species, some do occur in more acid soils but, as the soil pH drops, they are increasingly replaced by species of another, smaller (< 3 cm/1.2 in) type of annelid worm, the so-called potworms (Enchytraeidae) (Hågvar and Abrahamsen, 1980). They mostly replace earthworms in cool, moist acidic soils with a high content of organic matter. There are some other invertebrate types that also appear to do better in acid soils, for example oribatid mites (Cryptostigmata), as already discussed (Mueller et al., 2016). Low pH levels (say < pH 5) are unlikely to be encountered in most domestic compost heaps, with the possible exception of one very dominated by inputs of conifer twigs and needles. As a general rule, for compost that will be used in the garden a pH between pH 6 and pH 7 will work well and is likely to make life easy for a wide range of decomposer invertebrates.

Secondary metabolites

Plants can generate a bewildering array of so-called chemical 'secondary metabolites' in their tissues. There are many potential effects of these in compost (Chomel et al., 2016). Tannin polymers were discussed earlier in relation to invertebrate decomposers. Some studies have reported negative impacts on invertebrates of toxic alkaloids (for example, nicotine) produced by plants or their symbiotic fungi. On the other hand, we know that alkaloids are short lived and would not expect them to persist in senescing leaves. So, in reality it is hard to gauge the impact of this chemical complexity in the context of garden compost; just note that the potential for an effect is there.

We can finish this section by re-emphasizing the potential for interaction between the various factors discussed. Higher temperatures may mean less moisture. Less moisture means less decomposition and a lower requirement for oxygen. Too much moisture in the compost means less oxygen will be able to diffuse into the compost. Lower temperatures mean less decomposition and less oxygen required. Too much moisture (rainfall) seeping through the compost may wash out dissolved materials (called leaching) and result in changes of pH. And so on.

GARDEN ORGANIC WASTE IS NOT A NUISANCE BUT A RESOURCE

10 / THE LAST WORD...

A sign seen in the Botanic Garden in Singapore is perhaps both a good starting point as well as an end point to the story of decomposition and compost. It reads:

Oh my!
Look at all these leaves!
Fallen leaves, also known as leaf litter, are applied to the tree root zones in many areas of the Gardens to improve plant growth. Leaf litter carries a large amount of beneficial microorganisms which aid in the decomposition of organic matter. They also keep our plant roots moist during a hot day.

Leaves are the number one ingredient in nutrient recycling so remember to keep them in your planting beds too!

This represents leaf litter not as a nuisance, but as a resource to be used for the benefit of plants in the garden. Leaves can be left to rot down naturally where they fall or gathered up into compost heaps and bins. Either way, this may lead visitors to ask questions about the seemingly mysterious process of decomposition. How does nature dispose of its organic plant waste so effectively? Of course, it isn't just disposal - we have seen that it is all about recycling of materials, which then makes the natural world go round.

We have discussed the many and varied processes and factors leading to decomposition of organic matter, whether naturally or in compost heaps and bins, and have attempted to explain them on the basis of what is known. Clearly, scientific consensus is still lacking in some areas. But it is possible to have a broadly scientific view of decomposition - to help make composting 'work'. The end product should be a well-structured and nutritionally useful soil-conditioning or potting compost. There are many practical guides in book form, articles or web pages to compost making that show how this can be achieved. If you subscribe to the general ethos of recycling, then this is one way in which you can create more of a 'closed loop' in terms of nutrient cycling in the garden and improve your soil health too. These ideas are increasingly being promoted in agriculture/horticulture under headings such as regenerative or circular agriculture. The focus is on using 'waste' biomass and so minimizing inputs by promoting closed-loop nutrient cycling, soil regeneration and generally reducing environmental impacts. Making compost is one small step towards circular or regenerative gardening: reducing the environmental impact of your gardening activities by relying less on external resources. Even if you don't have the capacity to recycle all your green garden and kitchen waste, you will contribute to reducing the energy cost of transporting green waste by processing at least part of your organic material yourself.

Beyond thoughts of recycling and making good compost, this book has been for the curious – those of you who might want to delve into what really does go on in a compost heap. The word ecosystem, which came from the science of ecology, has been adopted in fields as varied as economics and information technology. But the compost heap is a real ecosystem, in the original meaning of the word. A pile of compost may seem unexciting, but you now have an insight into the complexity of the system. In any ecological system, measuring all the parameters involved and trying to model its functioning is a herculean task. In reality, ecologists can understand the general functioning of a particular ecosystem, but it is too variable and complex to predict fully. The statistician George Box once wrote that all models are wrong. This has since been expanded to 'All models are wrong, but some are useful.' So, there can be useful generalizations about the important factors that drive compost ecosystems and determine the direction of travel. What has been covered here is not a full description of compost ecology, but an outline of useful principles that can be applied to make composting work. The interested gardener, having read this book, can think about the variety and layering of organic matter loaded into the compost heap. This covers a lot of the points discussed with reference to nutrients and the physical environment in the compost, such as nitrogen, temperature, moisture and aeration – major factors in the rate of decomposition. Beyond that, the compost heap sleuth requires a magnifying glass and will need to get close and personal with their pile. When we go compost dipping, a sample taken from the compost and spread on a white surface begins to reveal some of its invertebrate riches. The microbes are more difficult, but a microscope may show fungal hyphae and possibly fungi sporulating on a substrate. And there will be a surprising variety of invertebrates. Maybe at first a single mite is visible – a small eight-legged relative of spiders. But as you look closer (higher magnification), it will become clear that there isn't just one type of mite, but potentially dozens. You'll recall the comparison made between a coral reef and life in a compost heap. A compost heap may be nothing like as colourful, yet (on a smaller scale) it shares a similar degree of variety of species and ecological complexity. And whether at home or at school, it is a good starting point for teaching children (for example, see Donze and Wong 2018) and adults that this is not 'just a boring compost heap'. It is about ecosystems and recycling at their most interesting and engaging.

THE GROUND'S GENEROSITY TAKES IN OUR COMPOST AND GIVES BEAUTY

—

Attributed to Rumi

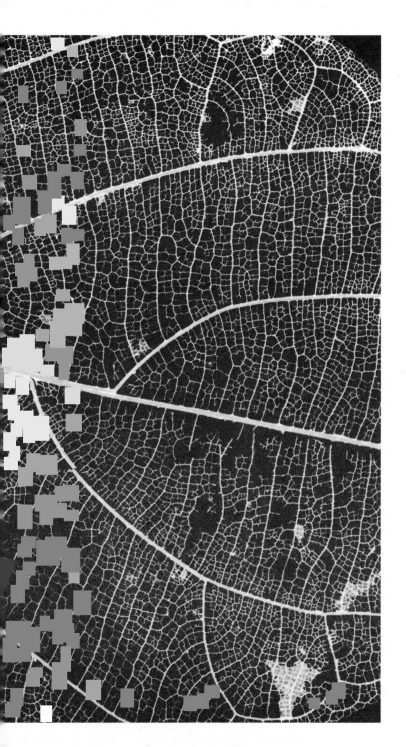

GLOSSARY

Definitions and explanations given here focus on terminology as used in the book.

Acari - mites are eight-legged invertebrates in the class Arachnida (spiders and mites).

Actinomycetes - gram-positive, filamentous bacteria that thrive in lower- or no-oxygen conditions (anaerobic). They are known to be a source of antibiotics.

Aerobic - cellular respiration (metabolism) that requires oxygen to break down (oxidize) a substrate and release energy.

Alkaloids - small organic nitrogen-containing secondary metabolites of plants, many of which are potentially toxic to animals.

Ammonia - a nitrogen-containing product of decomposition that will dissolve in water or can escape into the atmosphere as a gas.

Amorphous - material without defined shape or form.

Anaerobic - cellular respiration that can partially break down a substrate in the absence of oxygen, but has a lower energy yield than aerobic respiration.

Annelids - segmented worms, including earthworms (Lumbricidae) and potworms (Enchytraeidae).

Antibiotics - secondary metabolites released by cells of fungi or bacteria/actinomycetes into their immediate environment. They may have antimicrobial effects against other species of microorganisms.

Ascarosides - small signalling molecules released by roundworms (nematodes) that influence the behaviour of others of the same or different species.

Assimilation - absorption of nutrients (for example, nitrogen-containing molecules or ions) by organisms and leading to their incorporation in new organic molecules.

Astigmata - one of the four non-parasitic orders of mites.

Available nutrients - soil/compost nutrients that are in a soluble form as small molecules or ions which can be taken up by plant roots. The opposite term, for nutrients that cannot be taken up by roots, is 'unavailable'.

Bacteria - primitive single-celled microbes (prokaryotes). Unlike more advanced cells (eukaryotes), there is no membrane-bound nucleus.

Base - any substance that reacts with and neutralizes acids.

Calcicole - a plant that tolerates a chalky soil.

Calcifuge - a plant that cannot thrive in a chalky soil and grows in neutral or acid soils.

Cellulase - a family of enzymes that facilitate the decomposition of cellulose.

Cellulose - the major structural carbohydrate (polysaccharide) component of plant cell walls. Long-chain (polymer) carbohydrate molecules of linked simple sugars (monomers).

Chitin - a fibrous polysaccharide material that structures the cell walls of fungi and contains nitrogen (also in some invertebrates).

Chitinase – a family of enzymes that facilitate the decomposition of chitin.

Classification (biological) – grouping living organisms, originally on the basis of similarity of form but now also using DNA barcoding.

Collagen – major structural protein in animal tissues made of nitrogen-containing amino acid sub-units.

Collembola – six-legged insect relatives, common name springtails. Part of the decomposer community.

Compost – decaying plant material that can be added to soil to improve its quality.

Cryptostigmata – one of four non-parasitic orders of mites (this group is also known as Oribatida).

Cyanobacteria – photosynthetic bacteria that require light. Also known as blue-green bacteria (incorrectly as blue-green algae).

Desiccation – drying out/loss of moisture from whole organisms or tissues or cells.

Detritivore – feeding on dead plant or animal organic remains (detritus).

Diffusion – natural movement of molecules through a medium such as still water or still air. Movement is generally slow through still air and much slower through still water.

Enchytraeids – small, segmented worms (potworms: Enchytraeidae), related to earthworms (Lumbricidae). Part of the decomposer community.

Enzymes – proteins acting as biological catalysts to speed up specific chemical reactions in a cell.

Food web – the complete and complex network of feeding relationships in a compost heap. What eats what.

Fungi – a separate kingdom (from animals, plants and bacteria) of multicellular filamentous organisms that are characterized by the polysaccharide chitin in their cell walls. Yeasts are also fungi, but single-celled.

Hemicellulose – non-structural carbohydrate (polysaccharide) component of plant cell walls.

Hexapods – invertebrates with six legs (may or may not have wings).

Humic substances – large, complex polymer molecules made up of repeating six-carbon ring-structure sub-units (monomers) that are formed during decomposition.

Immobilization (of nutrients) – nutrients (elements) that are locked up in large, often complex molecules in a living or non-living form and that are not available for use by plants.

Ions – an atom or molecule carrying a positive or negative charge. For example, ammonia in solution (NH_4OH) splits (dissociates) into NH_4^+ and OH- ions. Another ion is NO_3^- = nitrate ion.

Leaching – the washing out of soluble nutrients from soil or compost by water from rainfall.

Lignase – a family of enzymes that facilitate the decomposition of lignins.

Lignin – major structural material giving strength to woody plant tissues. Large polymer molecules made up of repeating six-carbon ring-structure sub-units (monomers).

Lignocellulose – collective term for the cell-wall components of woody tissues, containing cellulose, hemicellulose and lignin.

Lipids – basic building material of living organisms. Constructed from carbon, hydrogen and oxygen and insoluble in water. Found in cells and tissues, some of that may store lipids as an energy reserve.

Litter (leaf) – fallen dead leaves from a parent plant.

Macrofauna – invertebrates (found in compost) > 2 mm/0.08 in in length.

Macronutrients – nutrient elements that plants need to take up in large amounts, primarily from soil and through the root system.

Mesofauna – medium-sized (0.1-2 mm/0.004-0.08 in) invertebrates found in compost.

Mesostigmata – one of four non-parasitic orders of mites.

Metabolism – all the basic biochemical processes that function continuously in organisms in cells/tissues to sustain life.

Microbiome – an assemblage of different microorganisms forming a microbial community at a particular location (for example, a leaf or root surface).

Microbiota – an assemblage of different microorganisms forming a microbial community.

Microfauna – small-sized (< 0.1 mm/0.004 in) invertebrates found in compost.

Micronutrients – nutrient elements (trace elements) that plants need to take up in small quantities, primarily from soil and through the root system.

Mineralization – the conversion of a nutrient element by decomposition from an organic form to a soluble inorganic form that makes it available for uptake by plants.

Mitochondria – organelles present in the cytoplasm of a plant, animal or fungal cell. They are the location for cellular respiration, where a sugar is broken down to release energy. This energy is first stored in adenosine triphosphate (ATP) molecules, then used to drive biochemical processes in the cell.

Molecule – a unit of two or more atoms held together by a chemical bond.

NPK – a compound synthetic fertilizer containing three macronutrient elements: nitrogen (N), phosphorus (P) and potassium (K).

Oxidation – in the context of a compost heap, this is the breakdown of larger organic molecules into smaller ones. Oxidation can be thought of as a process in which a substance gains oxygen, but the term now has a more technical meaning covering a chemical process in which a substance loses electrons.

Parasites – organisms living on or in a different host species, causing it some harm (for example, taking nutrients from the host).

Pathogens – disease-causing organisms.

pH – a scale with values from 0 - 14 for measuring the acidity or alkalinity (basicity) of a solution.

Phospholipids – a class of lipids that are a key component of cell membranes.

Phylloplane – the leaf surface, as a habitat for microorganisms.

Polymer – large or very large molecules of many repeating sub-units (monomers).

Polysaccharides – long-chain (polymer) carbohydrate molecules of linked simple sugars or derivatives of sugars (monomers).

Proteins – basic building materials of living organisms. Constructed from strings of different nitrogen-containing amino acids and fulfilling many roles, such as structural components and enzymes.

Resorption – the autumn transfer of nutrients from a dying (senescent) leaf to the parent plant.

Rhizosphere – a narrow region of soil surrounding a plant root influenced by root secretions and associated microorganisms (the root microbiome).

Saprophyte – a plant or microorganism (bacteria or fungi) that takes its nutrients from decaying dead organic matter.

Saproxylic – insects that spend much of their life cycle developing in dead wood and consuming both wood and associated fungi.

Sclerenchyma – supporting plant tissue with lignin-strengthened cell walls.

Secondary metabolites – in plants these are generally small molecules produced by individual species that are not essential for basic metabolic activities related to plant growth, development and reproduction.

Senescence (leaf) – gradual and programmed death and shedding of leaves, coupled with withdrawal of nutrients back to the parent plant.

Sporangium (fungal) – a stalked 'capsule' containing fungal spores.

Succession (compost) - a sequence of microbial or invertebrate species that colonize compost as it ages.

Sugar - carbohydrate molecules of carbon, hydrogen and oxygen. Soluble simple sugars include glucose (monosaccharide) and sucrose (disaccharide); sucrose has two monosaccharide units linked by a glycosidic bond.

Symbiosis - a term that covers a close and long-term living together of two different species. Covers both mutualism (living together for mutual benefit) and parasitism but is often used (incorrectly) as a term equivalent only to mutualism.

Tannins - large and complex polymer molecules made up of repeating six-carbon ring-structure sub-units (monomers).

Taxon - a taxonomic group that can be at any level in the classification, such as kingdom, class, family or genus.

Taxonomy - the science of classifying organisms into groups based on form and/or DNA analysis (classification).

Thermophile - a microorganism that has the capacity to thrive at temperatures above 41°C/105.8°F.

Translocation - in plant physiology this term refers to the movement of nutrients or other molecules from one part of a plant to another.

REFERENCES

Aerts, R. 1996. Nutrient resorption from senescing leaves of perennials: Are there general patterns? *Journal of Ecology*, Volume 84, pp. 597-608.

Ali, M. and Mindari, W. 2015. Effect of humic acid on soil chemical and physical characteristics of embankment. MATEC web of Conferences, 2016 - matec-conferences.org. BISSTECH 2015. DOI: 10.1051/matecconf/20165801028.

Anderson, J. 1983. Life in the soil is a ferment of little rotters. *New Scientist*, Volume 100, pp. 29-37.

Ashton, I. W. et al. 2005. Invasive species accelerate decomposition and litter nitrogen loss in a mixed deciduous forest. *Ecological Applications*, Volume 15, pp. 1,263-1,272.

Bardgett, R. D. and van der Putten, W. H. 2014. Belowground biodiversity and ecosystem functioning. *Nature*, Volume 515, pp. 505-511.

Brackin, R. et al. 2017. Soil biological health - What is it and how can we improve it? *Proceedings of the Australian Society of Sugar Cane Technologists*, Volume 39, pp. 141-154.

Brussaard, L. et al. 2007. Soil fauna and soil function in the baric of the food web. *Pedobiologia*, Volume 50, pp. 447-462.

Cáceres, R. et al. 2018. Nitrification within composting: A review. *Waste Management*, Volume 72, pp. 119-137.

Cepelewicz, J. 2021. Radioactivity may fuel life deep underground and inside other worlds. [Online] Available at: Radioactivity May Fuel Life Deep Underground and Inside Other Worlds | Quanta Magazine [Accessed 22 February 2022].

Chomel, M. et al. 2016. Plant secondary metabolites: A key driver of litter decomposition and soil nutrient cycling. *Journal of Ecology*, Volume 104, pp. 1,527–1,541.

Cornelissen, J. 1996. An experimental comparison of leaf decomposition rates in a wide range of temperate plant species and types. *Journal of Ecology*, Volume 84, pp. 573–582.

Cornell Waste Management Institute, 1996a. Cornell Composting - Compost Chemistry, Ithaca: Cornell Waste Management Institute.

Cornell Waste Management Institute, 1996b. Cornell Composting - Calculating the oxygen diffusion coefficient in water, Ilthaca: Cornell Waste Management Institute.

Cornwell, W. K. et al. 2008. Plant species traits are the predominant control on litter decomposition rates within biomes worldwide. *Ecology Letters*, Volume 11, pp. 1,065–1,071.

Crowther, T. W. et al. 2011. Species-specific effects of soil fauna on fungal foraging and decomposition. *Oecologia*, Volume 167, pp. 535–545.

Darwin, C. 1888. *The Formation of Vegetable Mould Through the Action of Worms with Observations of their Habits*. London: John Murray. (Note: first edition 1881.)

Davies, J. 2006. Are antibiotics naturally antibiotics? *Journal of Industrial Microbiology and Biotechnology*, Volume 33, pp. 496–499.

Digel, C. et al. 2014. Unravelling the complex structure of forest soil food webs: Higher omnivory and more trophic levels. *Oikos*, Volume 123, pp. 1,157–1,172.

Donze, J. and Wong, S.S. 2018. Where did the leaves go? *Science Scope*, Volume 42, No. 2 (Earth Systems), pp. 94–102.

Fierer, N. and Jackson, R. B. 2006. The diversity and biogeography of soil bacterial communities. *Proceedings of the National Academy of Sciences* U.S.A., Volume 103, pp. 626-631.

Frouz, J. 2018. Effects of soil macro- and mesofauna on litter decomposition and soil organic matter stabilization. *Geoderma*, Volume 332, pp. 161-172.

Frouz, J. et al. 2011. Effect of soil invertebrates on the formation of humic substances under laboratory conditions. *Eurasian Soil Science*, Volume 8, pp. 973-977.

Fujii, K. et al. 2012. Isolation and characterization of aerobic microorganisms with cellulolytic activity in the gut of endogeic earthworms. *International Microbiology*, Volume 15, pp. 121-130.

Geisseler, D. et al. 2010. Pathways of nitrogen utilization by soil microorganisms - A review. *Soil Biology & Biochemistry*, Volume 42, pp. 2,058-2,067.

Guerreiro, M. A. et al. 2018. Transient leaf endophytes are the most active fungi in 1-year-old beech leaf litter. *Fungal Diversity*, Volume 89, pp. 237-251.

Hagen-Thorn, A. et al. 2006. Autumn nutrient resorption and losses in four deciduous forest tree species. *Forest Ecology and Management*, Volume 228, pp. 33-39.

Hågvar, S. and Abrahamsen, G. 1980. Colonisation by Enchytraeidae, Collembola and Acari in sterile soils samples with adjusted pH levels. *Oikos*, Volume 34, pp. 245-258.

Heal, O. W. et al. 1997. Plant litter quality and decomposition: An historical overview. In G. Cadisch and K. Giller, eds, *Driven by Nature: Plant Litter Quality and Decomposition*. CAB International, pp. 3-30.

Hedlund, K. et al. 1995. Fungal odour discrimination in two sympatric species of fungivorous collembolans. *Functional Ecology*, Volume 9, pp. 869-875.

Horwath, W. R. et al. 1995. Mechanisms regulating composting of high carbon to nitrogen ratio grass straw. *Compost Science & Utilization*, Volume 3, pp. 22-30.

Hsueh, Y. et al. 2013. Nematode-trapping fungi eavesdrop on nematode pheromones. *Current Biology*, Volume 23, pp. 83-86.

Jones, G. L. et al. 2019. Litter of the invasive shrub *Rhododendron ponticum* (Ericaceae) modifies the decomposition rate of native woodland litter. *Ecological Indicators*, Volume 107, 105597.

Liao, C. et al. 2007. Altered ecosystem carbon and nitrogen cycles by plant invasion: A meta-analysis. *New Phytologist*, Volume 177, pp. 706-714.

López-Mondéjar, R. B. et al. 2018. Decomposer food web in a deciduous forest shows high share of generalist microorganisms and importance of microbial biomass recycling. *ISME Journal*, Volume 12, pp. 1,768-1,778.

Maraun, M. et al. 2003. Adding to 'the enigma of soil animal diversity': Fungal feeders and saprophagous soil invertebrates prefer similar food substrates. *European Journal of Soil Biology*, Volume 39, pp. 85-95.

Meentemeyer, V. 1978. Macroclimate and lignin control of litter decomposition rates. *Ecology*, Volume 59, pp. 465-472.

Miyashita, K. et al. 1982. Actinomycetes occurring in soil applied with compost. *Soil Science and Plant Nutrition*, Volume 28, pp. 303-313.

Mueller, K. E. et al. 2016. Light, earthworms, and soil resources as predictors of diversity of 10 soil invertebrate groups across monocultures of 14 tree species. *Soil Biology & Biochemistry*, Volume 92, pp. 184-198.

Phillips, H. R. P. et al. 2019. Global distribution of earthworm diversity. *Science*, Volume 366, pp. 480-485.

Polishchuk, L. V. and Blanchard, J. L. 2019. Uniting discoveries of abundance-size distributions from soils and seas. *Trends in Ecology and Evolution*. Volume 34, pp. 2-5.

Reich, P. B. et al. 2005. Linking litter calcium, earthworms and soil properties: A common garden test with 14 tree species. *Ecology Letters*, Volume 8, pp. 811-818.

Rousk, J. et al. 2011. Fungal and bacterial growth responses to N fertilization and pH in the 150-year 'Park Grass' UK grassland experiment. *FEMS Microbiology Ecology*, Volume 76, pp. 89-99.

Ruppert, E. E. et al. 2004. *Invertebrate Zoology: A Functional Evolutionary Approach*. Belmont, CA: Brooks/Cole.

Ryckeboer, J. et al. 2003. Microbiological aspects of biowaste during composting in a monitored compost bin. *Journal of Applied Microbiology*, Volume 94, pp. 127-137.

Sanger, L. J. et al. 1998. Variability in the quality and potential decomposability of Pinus sylvestris litter from sites with different soil characteristics: Acid detergent fibre (ADF) and carbohydrate signatures. *Soil Biology & Biochemistry*, Volume 30, pp. 455-461.

Satchell, J. E. 1974. Litter - Interface of animate/inanimate matter. InG. Pugh and C. Dickinson, eds, *Biology of Plant Litter Decomposition*. London: Academic Press, pp. 14-44.

Schlatte, G. et al. 1998. Do soil microarthropods influence microbial biomass and activity in spruce forest litter? *Pedobiologia*, Volume 42, pp. 205-214.

Schloss, P. D. and Handelsman, J. 2006. Toward a census of bacteria in soil. *PLoS Comput.Bio.*, Volume 2: e92. DOI: 10.1371/journal.pcbi.0020092.

Seastedt, T. R. 1984. The role of microarthropods in decomposition and mineralization processes. *Annual Review of Entomology*, Volume 29, pp. 25-46.

Sechi, V. et al. 2014. Collembola feeding habits and niche specialization in agricultural grasslands of different composition. *Soil Biology & Biochemistry*, Volume 74, pp. 31-38.

Shen, Q. et al. 2016. Testing an alternative method for estimating the length of fungal hyphae using photomicrography and image processing. *PloS ONE*, Volume 11: e0157017. DOI: 10.1371/journal.pone.0157017.

Slade, E. M. and Riutta, T. 2012. Interacting effects of leaf litter species and macrofauna on decomposition in different litter environments. *Basic and Applied Ecology*, Volume 13, pp. 423-431.

Smith, S. R. and Jasim, S. 2009. Small-scale home composting of biodegradable household waste: Overview of key results from a 3-year research programme in West London. *Waste Management & Research*, Volume 27, pp. 941-950.

Stevenson, F. J. 1994. *Humus Chemistry: Genesis, Composition, Reactions* (2nd edition). New York: Wiley.

Sun, Z. et al. 1997. Earthworm as a potential protein source. *Ecology of Food and Nutrition*, Volume 36, pp. 221-236.

Törmälä, T. 1979. Numbers and biomass of soil invertebrates in a reserved field in central Finland. *Journal of the Scientific Agricultural Society of Finland*, Volume 51, pp. 172-187.

Tournier, V. et al. 2020. An engineered PET depolymerase to break down and recycle plastic bottles. *Nature*, Volume 580, pp. 216-219.

Van den Hoogen, J. et al. 2019. Soil nematode abundance and functional group composition at a global scale. *Nature*, Volume 572, pp. 194-198.

Vaverková, M. et al. 2014. How do degradable/biodegradable plastic materials decompose in home composting environment? *Journal of Ecological Engineering*, Volume 15, pp. 82-89.

Vivelo, S. and Bhatnagar, J. M. 2019. An evolutionary signal to fungal succession during plant litter decay. *FEMS Microbiology Ecology*, Volume 95, pp. 1-11.

Voříšková, J. and Baldrian, P. 2013. Fungal community on decomposing leaf litter undergoes rapid successional changes. *ISME Journal*, Volume 7, pp. 477-486.

Wang, X., et al. 2007. Rapid and automated enumeration of viable bacteria in compost using a micro-colony auto counting system. *Journal of Microbiological Methods*, Volume 71, pp. 1-6.

Whitman, W. B. et al. 1998. Prokaryotes: The unseen majority. *Proceedings of the National Academy of Sciences*, Volume 95, pp. 6,578-6,583.

Zhang, K. L. et al. 2017. The effects of combinations of biochar, lime, and organic fertilizer on nitrification and nitrifiers. *Biology and Fertility of Soils*, Volume 53, pp. 77-87.

Zimmer, M. et al. 2005. Do woodlice and earthworms interact synergistically in leaf litter decomposition. *Functional Ecology*, Volume 19, pp. 7-16.

INDEX

ACKNOWLEDGEMENTS

This short book was largely written during the first UK 'Covid lockdown'. The ebb and flow of scientific evidence and debate about the workings of soil and compost was a good distraction from the pandemic! That is the nature of science – endlessly fascinating and evolving. So, thanks go to all those researchers who produced research papers which kept me busy during that time. Thanks also go to the many students I taught ecology over the years – they also helped to keep me enthusiastic about the many varied aspects of ecology, including the workings of soil.

The initial book proposal was taken up by Anna Sanderson of Pimpernel Press in London, who has also ably overseen the publication process. I'm grateful that my writing efforts may introduce a wider audience to the hidden world of compost. Thanks go to my wife, Christine, for insightful comments on the initial text of the book. Subsequent, very helpful contributions came from my editor Monica Hope. Sarah Pyke and Becky Clarke had the difficult job of converting my sketchy figures into the finished articles – thank you.

PERMISSIONS

The figures in the book were generally devised by me or, in some cases, as my versions of frequently published generic figures. The illustrations of invertebrates in Figures 8,9 and 10 were taken from Kevan, D.K. McE. 1955. *Soil Zoology*. London: Butterworth. I am grateful for permission to use these, which was granted by Elsevier Publishers, as far as they were able to establish the copyright ownership of these illustrations. Figure 6 was very useful and available copyright free from the journal *FEMS Microbiology Ecology* (Vivelo, S. and Bhatnagar, J.M. 2019, Volume 95, pp. 1-11).